Practical Manual on Fundamentals of Entomology

About the Authors

Dr. Biwash Gurung is currently working as an Assistant Professor in School of Agricultural Sciences (ICAR Accredited) G.D Goenka University Gurugram, Haryana. He has obtained his Ph.D Degree in Agricultural Entomology from Uttar Banga Krishi Viswavidyalaya, (North Bengal Agricultural University), West Bengal. He has qualified ICAR -NET 2018. He holds one year eight months of teaching experience at ITM University, Gwalior, before joining G.D Goenka University. He has published more than 16 research paper, 2 short communications, 2 Book chapters, 6 popular articles and 1 paper at Book of proceedings. Dr. Gurung has reported "New prey record of giant ladybird beetle Anisolemnia dilatata (Fabricius) (Coccinellidae: Coleoptera feeding on Som plant Aphid Aiceona sp". Dr.Gurung has delivered Plenary lecture at 11th International Conference on Technology, Innovation, and Management for Sustainable Development (TIMS- 2023). He has also attended workshops, conferences and symposium at National and International level.

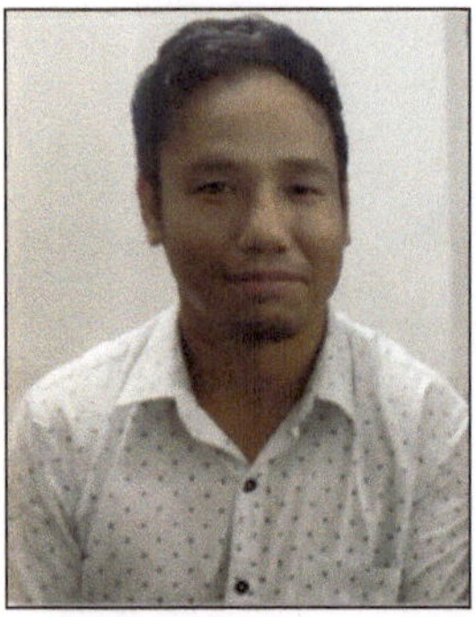

Dr. Bikash Subba serves as an Assistant Professor of Agricultural Entomology at the School of Agricultural Science, GD Goenka University. Prior to this role, he held positions as a lecturer at Sikkim University and ITM University. Dr. Subba specializes in the fields of Biological Control of Pests and Diseases, Insect Ecology, Insect Taxonomy, Integrated Pest Management, Crop Protection Entomology, and Pollination with a focus on insect pollinators. He earned his M.Sc. and Ph.D in Agricultural Entomology from Uttar Banga Krishi Viswavidyalaya. Throughout his career, Dr. Subba has demonstrated a commitment to research, having published 15 research papers in renowned journals and contributed to two book chapters. Additionally, he has presented papers at various national and international conferences. Dr. Subba is not only dedicated to academic pursuits but also actively engages in practical initiatives. He has organized numerous workshops aimed at imparting technical knowledge on organic farming to farmers in Sikkim. His efforts contribute to the dissemination of valuable information and the promotion of sustainable agricultural practices.

Practical Manual on Fundamentals of Entomology

Biwash Gurung M.Sc (Ag) Ph.D
Assistant Professor
Department of Agricultural Entomology
School of Agricultural Sciences
G.D Goenka University, Sohna, Gurugram-122 103
Haryana, India

Bikash Subba M.Sc (Ag) Ph.D
Assistant Professor
Department of Agricultural Entomology
School of Agricultural Sciences
G.D Goenka University, Sohna, Gurugram-122 103
Haryana, India

PMW

PMW
FU 75, Pitam Pura
New Delhi 110034
Mob. 91-8447075807
Email: publishmywork23@gmail.com
Web: www.publish-mywork.com

Print ISBN: 978-93-61341-55-7

ebook ISBN: 978-93-61345-39-5

Prof. (Dr.) Shiv Singh Tomar
Dean, School of Agricultural Sciences
G D Goenka University
Sohna-Gurgaon Road (Haryana)-Pincode: 122103, India

Foreword

The study of insects, their biology, and their impact on ecosystems is both fascinating and integral to understanding the complex web of life on our planet. As we explore into the diverse and captivating sphere of entomology, I am pleased to present this **Practical Manual on Fundamentals of Entomology**. This manual has been scrupulously prepared to serve as a comprehensive guide for students, researchers, and enthusiasts eager to explore the fundamentals of entomology in a practical and hands-on manner. The intricate world of insects holds the keys to unlocking mysteries that range from ecological balance to agricultural sustainability, making this manual an essential companion for anyone venturing into the field.

The manual is structured to provide a holistic understanding of entomology, covering a range of topics from the morphology, anatomy and physiology of insects to their classification, behavior, and ecological roles. The emphasis on practical exercises ensures that learners not only grasp theoretical concepts but also gain valuable experience in the field and laboratory settings. Every chapter is structured to help with a step-by-step learning process, beginning with basic ideas and gradually advancing to more complex topics. The goal of adding detailed picture and practical exercises is to improve the educational process and make entomology a fascinating and rewarding subject.

I would like to express my gratitude to the dedicated faculties of G.D, Goenka university, whose expertise has contributed for the preparation of this manual. Their commitment to the field of entomology and passion for knowledge-sharing have played a crucial role in shaping the content and ensuring its relevance to both beginners and those with a more advanced understanding of the subject.

With each new discovery and technological development, the study of entomology is constantly evolving, and this handbook provides a solid basis for grasping the essential concepts as entomological research. I hope that readers would approach the practical tasks with a sense of wonder and excitement since they will surely obtain deeper insights into the world of insects through firsthand encounters.

I hope this book will serve as a valuable resource in your journey to unravel the secrets of these remarkable creatures. May your exploration of entomology be filled with curiosity, discovery, and a profound appreciation for the intricate tapestry of life that insects contribute to our planet.

With Warms Regards

Prof. (Dr.) Shiv Singh Tomar
Dean, School of Agricultural Sciences
G D Goenka University,
Sohna-Gurgaon Road (Haryana)-India

Preface

Welcome to the fascinating world of entomology – the study of insects. Insects are among the most diverse and abundant organisms on Earth, playing crucial roles in ecosystems, agriculture, and human health. This practical manual has been crafted to serve as your guide into the fundamentals of entomology, providing a hands-on approach to understanding the morphology, behavior, and classification of these remarkable creatures.

Entomology is a multidisciplinary science that encompasses aspects of biology, ecology, physiology, and taxonomy. The **"Practical Manual on Fundamentals of Entomology"** is designed to cater to the needs of students, researchers, and enthusiasts who seek a comprehensive and practical resource to explore the world of insects. Whether you are a beginner or someone with a background in the biological sciences, this manual aims to enhance your understanding and appreciation of the intricate lives of insects.

The goals of this guidebook are to present basic insect collecting and preservation techniques, alterations to their appendages, identification features, external features of insects and spraying techniques and pesticide formulations.

A collage of material and text from different sources has been used to prepare this text book. The authors acknowledge their indebtedness to authors of books, manuals and internet media from which most of the materials have been drawn. In most of cases, it has obtained permission of reproduction, for which the authors and the publisher offer their apologies

The authors express their sincere gratitude to the Dean School of Agricultural Sciences Prof. (Dr.) S. S Tomar, G.D Goenka University, and other colleagues for their valuable suggestions in improving the quality from time to time during preparation of the practical manual. The authors highly acknowledged to our all well wishers for their direct-indirect cooperation.

Authors

Contents

1

To Study the Methods of Collection and Preservation of Insects Including Immature Stages

Insect collection is a source of recreation for many people and maybe a hobby for those who are interested in studying insects. Methods of collection and preservation of insects are the pre-requisite to studying the insects and their various internal and external organs. After collection, it becomes imperative to keep and preserve the insect specimens intact and safe for a longer time to further study the characters or to develop the insect collection museum. Let‘s have a look and do the different types methods of collection, and devices used for the collection and preservation of insects including immature stages in this practical session.

Collection

The purposes of insect collection are:

- To know the occurrence of a particular insect in a particular locality.
- Taxonomic study of insects.
- Identification of insects.
- To study the bionomics and rearing of insects for various entomological experiments.
- To develop the insect repository etc.

1. Instruments required

2. Insect collection net
3. Aspirator (poster)
4. Vials (specimen tubes) of various sizes
5. Forceps and hand lens

6. Hairbrush
7. Insect killing bottle
8. Setting board
9. Insect store box
10. Insect rearing cage
11. Insect killing and preservation media

Methods of Insect Collection

Hand-picking

This method is suitable for catching large insects like beetles and grasshoppers. It is a very tedious (hard-working) method and not suitable for catching the biting and stinging-natured insects.

Aerial net or Butterfly Net

It is light in weight, and useful for catching active fliers like butterflies, moths, dragonflies, wasps, flies, etc. The net consists of three parts viz., loop or frame; handle and porous muslin clothing bags. The diameter of the hoop and the depth of the bag should be in the proportion of 1:2.

Sweep Net

It is heavier than the aerial net. It consists of a short handle, a large loop and a dense cloth bag. This is suitable for collecting leafhoppers, grasshoppers, and other small insects. The net is swept over vegetation.

Beating Tray

This method is suitable for collecting crawling insects and those, which rest on branches. A beating tray is held under a branch, which is then hit sharply with a stick.

Aspirator/Potters/Suction tube

It is a device to collect small insects into glass vials with no damage to the specimens. It is employed to suck in through a rubber tube small and minute insects that are already collected in the net or sitting on a wall or foliage and the bark of the tree. Usually, it is meant for catching more active insects. To prevent the entry of insects into the mouth, a small cloth piece is kept in between the glass and rubber tube.

Berlese (Tullgren) Funnel

Soil arthropods can be sorted out by this method. Debris including soil arthropods can be collected by using light as the source of heat in the Berlese funnel method.

Traps

Trapping is a method of collecting insects in the absence of a collector. This is the most common method or technique used by growers in an Integrated pest management program to catch the insects. There are many different types of traps used for the collection of insects. They are pheromone traps or sleeve traps, fruit fly traps, sticky traps, delta traps, water or wota traps, pitfall traps, wind pan traps, malaise traps and light traps.

Pheromone Traps

Synthetic sex pheromones are placed in the traps to attract male moths. The rubberized septa containing the pheromone lure are kept in the traps designed especially for this purpose and used in monitoring, mass trapping and mating disruption programs. Sticky traps, Water pan traps and funnel-type models are available for use in pheromone-based insect-pest control programs.

Yellow sticky traps: Aphids, whiteflies, and thrips prefer yellow color. Yellow color is painted on tin boxes and sticky material like castor oil/vaseline is smeared on the sticky material.

Probe trap: It is used by keeping them under grain surface to trap stored product pests.

Pitfall traps: Containers such as small plastic buckets, plant pots, glass jars, or jam tins are sunk into the ground to trap flightless, ground-living insects and arachnids, especially beetles (ground beetles), cockroaches, crickets, spiders, harvestmen and mites. The container should be placed in a hole with the upper rim flush with the ground surface. A killing agent and preservative, such as ethylene glycol, should be placed in traps that are not emptied daily. Radiating vanes, such as wooden planks, placed in the substrate will increase the effective area of the trap. Bait can be added to the trap to increase its effectiveness. The type of bait will depend on the specimens one wishes to catch.

Light traps: Light traps are mainly used for attracting moths & other night-flying insects which are attracted to the light. The insects are actively caught or encouraged to enter a trap. The simplest light trap consists of light on a cable hanging out in the field to attract the pests during the night. However, besides several species of moths, beetles, flies, and other insects, most of which are not pests, are also attracted to artificial light. So identification of pests and beneficial insects is of prime importance before any control operation is executed.

Mercury vapor lamp light trap: This trap is the basic model designed by Robinson (1952). This trap produces ultraviolet, blue and green radiation with little red. This is currently used for a wide range of noctuid and other nocturnal flying insects. a mercury lamp (125W)is fixed at the top of a funnel-shaped (or) trapezoid galvanized iron cone terminating in a jar containing dichlorphos soaked in cotton as an insecticide to kill the insect.

Killing Insects: Killing should be immediately after capture. Potassium cyanide (KCN), ethyl acetate, carbon tetrachloride and chloroform are commonly used for killing insects. KCN kills the insects quickly but deadly poisonous and must be handled with extreme care. Ethyl acetate kills the insects slowly and does not last long. But the dead insects remain in relaxed condition for a longer time without becoming brittle and stiff.

Pinching- In this method, the thorax is pressed between the thumb and index finger swiftly and with jerk. It needs constant practice. e.g. butterfly, grasshopper.

Injecting- Hypodermic injection of fluids.

Drowning- Larvae and insects without scales, hairs or powdery covering can be killed by submerging them in water. They die of Auto toxicity when excessive CO2 unable to escape from spiracles and collects in the trachea and tissues.

How to prepare a Killing Bottle/ Cyanide Bottle?

The steps involved in preparing the killing bottles are given below

- Take a wide-mouthed strong bottle or vial with a tight fitting lid.
- Place a layer of potassium cyanide granules/pellets (1/4-inch thickness) at the bottom of the bottle
- Cover it with a layer of dry plaster of Paris (1/4-inch thickness)
- Mix the plaster of Paris with enough water so that it will pour off from the end of a spoon. Pour a 1/2-inch layer of wet plaster of Paris over the dry layer.
- Tap the bottle lightly on the table to eliminate any bubbles in the bottle.
- Leave the lid off for a day to let the plaster dry in a well-ventilated room, completely away from direct sunlight.

- Keep a circular piece of filter or blotting paper on the top of the plaster of Paris and avoid condensation of water droplets on the side of the bottom (to check the sweating process).
- Lastly, the bottle should be tightly corked and labeled with the word ―Poison‖.
- In place of KCN, now a day Ethyl acetate is being used as the replacement in killing bottles.

Insect Collection Box

Storage of insects is done in the insect boxes, which is made up of wood (top and bottom could be of plywood) and lined on one (bottom) or both (roof also) sides with cork sheets covered with white paper. It is light in weight, moisture-proof and airtight. The general (common) size of the insect collection box is 45x30x15 cm.

Labeling

Specimen collected should be uniform in size and labeled properly on stiff paper or reference card. Labeling consists of the following notes i.e., **Host, Date, collector and Location.**

Setting or Stretching Boards

The setting is the method that wings antennae and Occasionally (Hymenopterans) spread legs in full display of their features. This method needs a setting or stretching board that has two sideboards separated by a groove. Both board and grooves are lined with a thin sheet of cork. The width of the groove varies according to the width of the insect's body.

Methods of Preservation

Protection of Insect Specimens

Collected Insects can be protected for a longer time in an insect collection box by putting the naphthalene balls on the corner side of the box.

Preservation of Insects

- **Temporary preservation**
  - **Permanent preservation**: Insects can be permanently preserved either dry, in fluid, or on microscope slides. Arachnids are always preserved in liquid or on microscope slides. The method of preservation depends on the type of arthropods. It can be done by the following methods-

1. **Dry preservation-** Insects that are to be preserved dry are best mounted in ways that facilitate study and permanent storage. Specimens should be mounted soon after killing, if possible while still soft.
2. **Liquid preservation-** It is done in 70 % ethyl alcohol + 4 % formaline solution.
3. **Soft-bodied insects and mealy bugs** can be preserved in a mixture of 4 parts 90 % ethanol and 1 part glacial acetic acid, **whereas thrips** can be preserved in a mixture of 9 parts 60 % ethanol and 1 part glacial acetic acid. It is very important to periodically check and top up containers of a liquid collection.
4. **Mounting on a microscopic slide-** Small specimens have to be mounted on microscope slides so that they can be studied under a compound microscope. These include groups such as thrips, aphids, parasitic wasps, scale insects, booklice, lice and mites. Insect and spider body parts (e.g. mouthparts and genitalia), and larvae often have to be slide-mounted. Microscope slide mounts may be temporary or permanent, but specimens maintained in collections require permanent mounts.

Bringing the Specimen Home or the Laboratory

Materials Required

Butterflies and other large-winged insects can be stored in folded protective paper envelopes. Most arthropod specimens can be conveniently stored between layers of absorbent paper. Paper envelopes (Newspaper, wax paper) can be used to keep the specimen and brought it to home or laboratory (having good absorbent quality) Cellophane and transparent plastics can also be used for this purpose.

Relaxing Container/Jar

Relaxing is the method/process of re-softening the insects. Relaxing container/ Jar –contains a layer of sand (5 cm thick) or any other absorbent materials (basal wood, pith, synthetic sponge) and a few drops of formalin or carbolic acid is added to prevent mold/fungal growth and then covered with filter paper. Cleaning- Dust, pollens and dirt can be removed with a camel hairbrush dipped in water mixed with detergent.

Preparing Insects for the Insect Collection Box

Insects longer than about 8 mm are usually mounted on pins pushed through the thorax. Insect pins are longer than ordinary pins and are made of stainless steel that does not rust. A No. 2 or No. 3 entomological pin is suitable for most insects, although those with delicate bodies may require a size No. 0 or No. 1.

Entomological pins- There are three general series of pins viz., **English pins:** Sold by weight, range of 18-30 mm in length and stout, used to pin Lepidopteran insects, which lies or are kept low in the box.

continental pins: Sold by 100s, Range 35 mm in length, 000,00,0 & 1-7 Nos., No. 2 & 3 are useful for general purpose, 38 mm (No. 8-10), 50 mm (No. 11-12), 000 is the thinnest pin and No. 12 is the thickest pin.

Minute pin: Minutest and finest pins, used for pinning the insects meant to stage, for a minute, softest, and fragile insects.

Pinning

It is the best and most common method to preserve hard-bodied insects. They will dry and remain in perfect condition on the pins for a long time without any further treatment. They are pinned vertically through the body. Depending upon the size of the insect pins have to be selected accordingly. The exact place of insertion of the pin varies among different groups of insects.

Insect groups Sketch diagram Pinning region

Insect groups	**Sketch diagram**	**Pinning region**
Grasshopper, crickets, Preying mantids and cockroaches	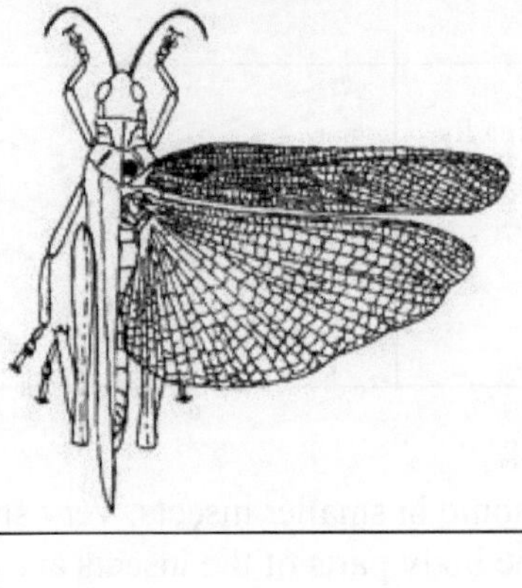	Pronotum

Insect groups	**Sketch diagram**	**Pinning region**
Bug (most of the Hemipterans)	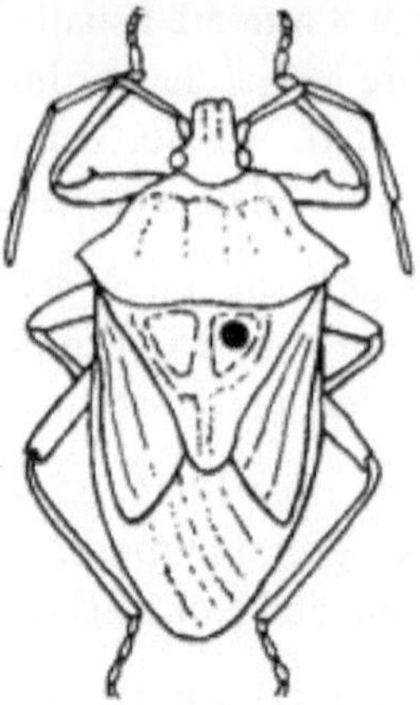	Scutellum
Beetles and weevils (Colepterans)	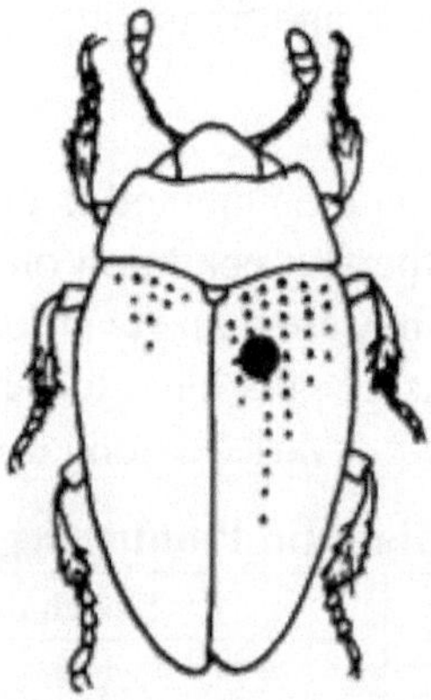	Right elytra
Dragonfly, damselfly, antlion, green lace wing fly, Moths, butterfly bees, wasp, ants and true flies etc.	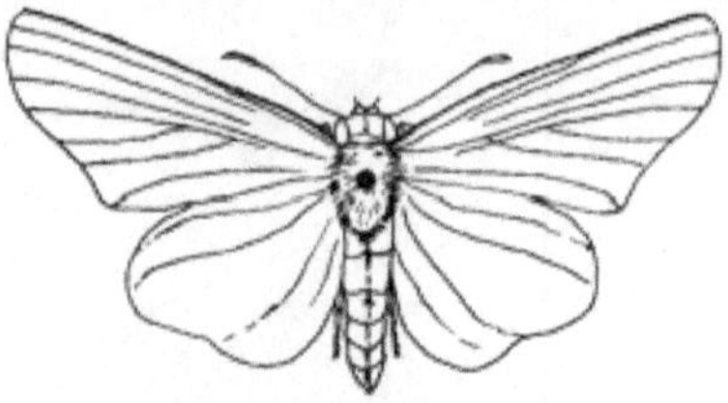	Thorax

Double Mounting

Pinning is troublesome in smaller insects. Very small insects cannot be pinned because most of the body parts of the insects are lost during pinning. For such insects, double mounting can be followed.

Staging

The stage is a narrow rectangular piece of cork or pith. The small insect is pinned correctly with a micro pin to the stage. Later the stage is pinned in the insect store box with a bigger pin.

Carding

A rectangular white card (**5x8 or 5x12mm**) may be used as a stage. On stage instead of pinning, the insect specimen is stuck on it by **using glue or adhesive**. After mounting the insect, the card is pinned in the box with a large pin.

Pointing/Gumming

The insect specimen is glued to a card cut into a triangle of 10 mm height and 5 mm base. Bend down the tip of the card to form a small surface on which the insect is stuck. Apply a drop of glue or adhesive by touching the point to the glue and to the thorax of the insects to be mounted.

Insect collection box

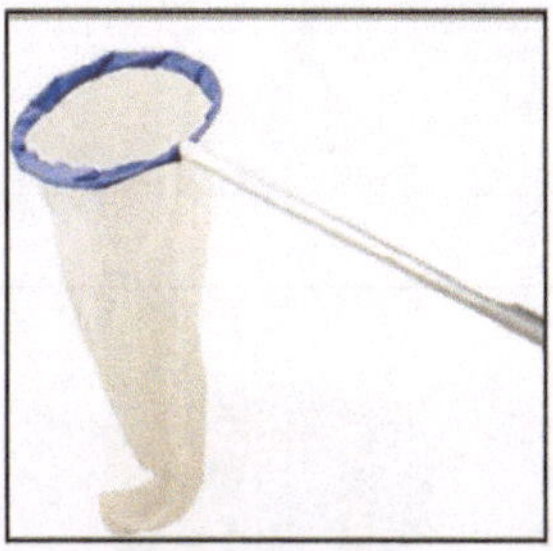

Hand Net

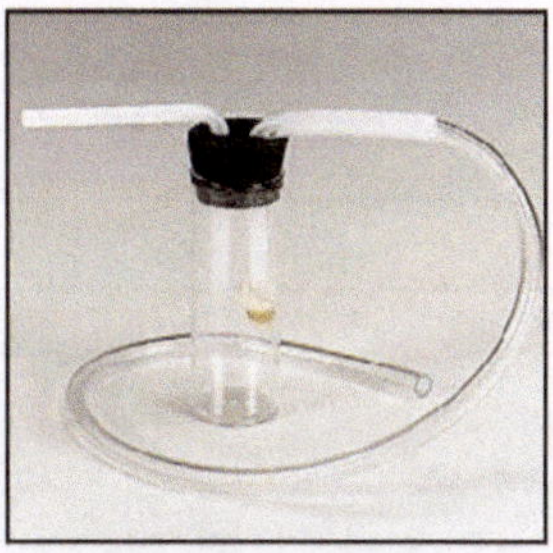

Aspirator

Hand Lens

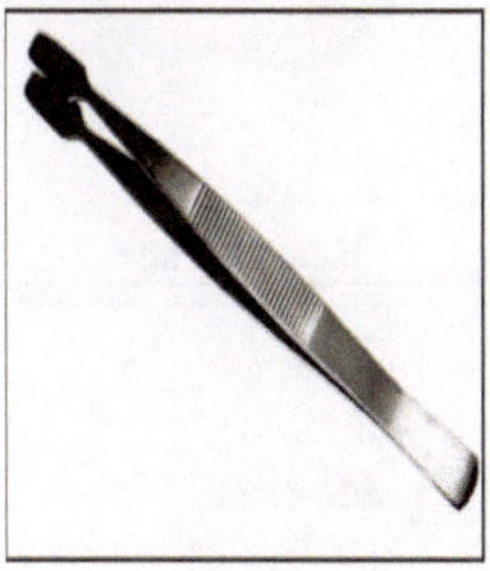

Forceps

Vial speciem tube

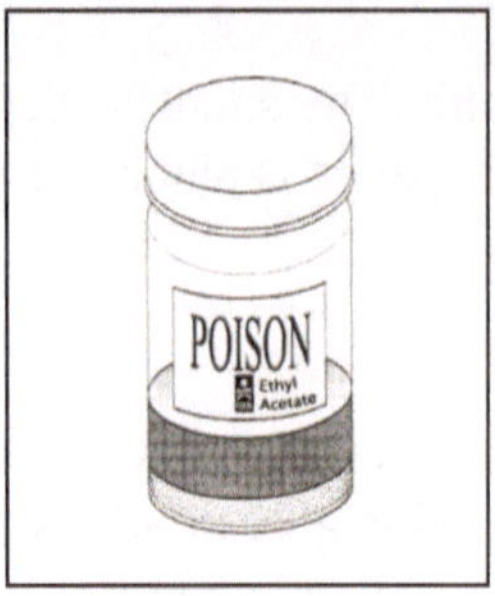

Killing bottle

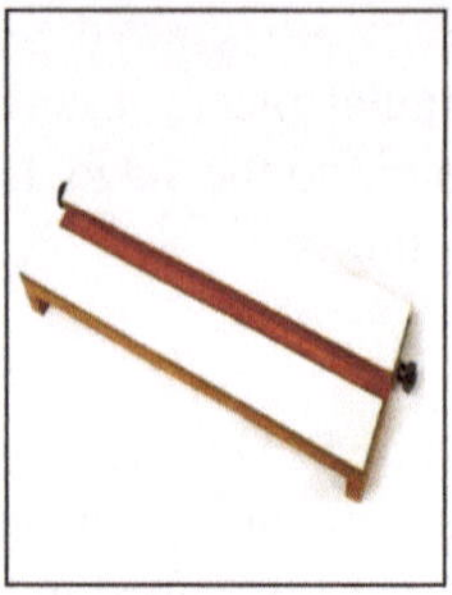

Setting Board

Display Box

Hair bush

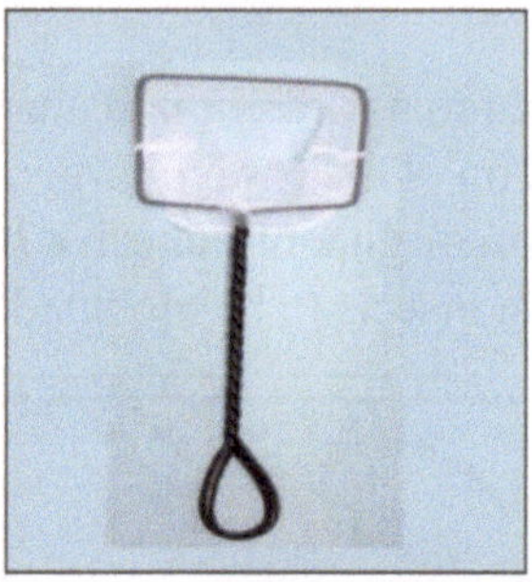

Pond Net

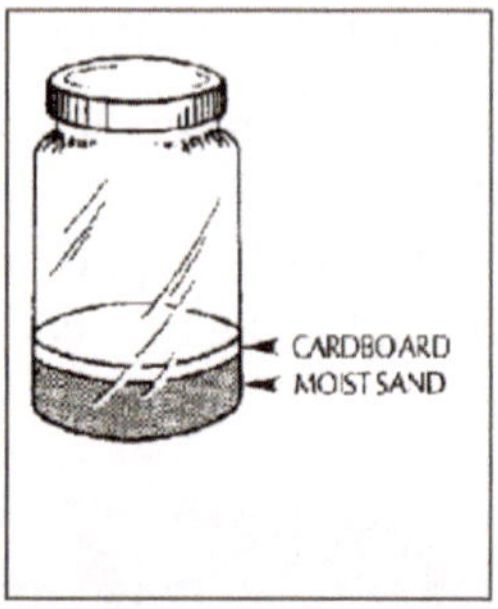

Relaxation chamber

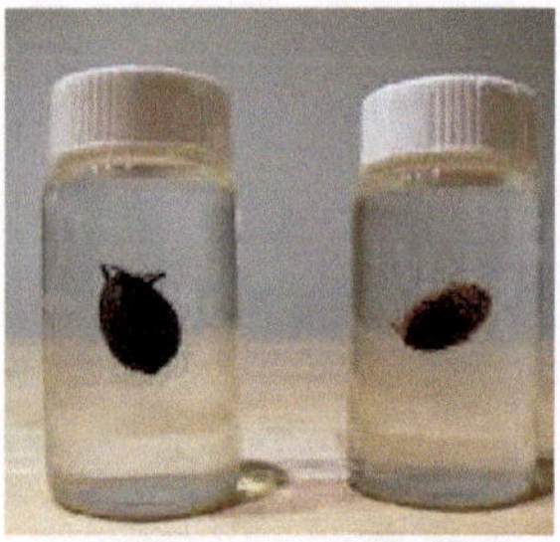

Liquid preservation

Rearing cage

Light trap

Pitfall trap

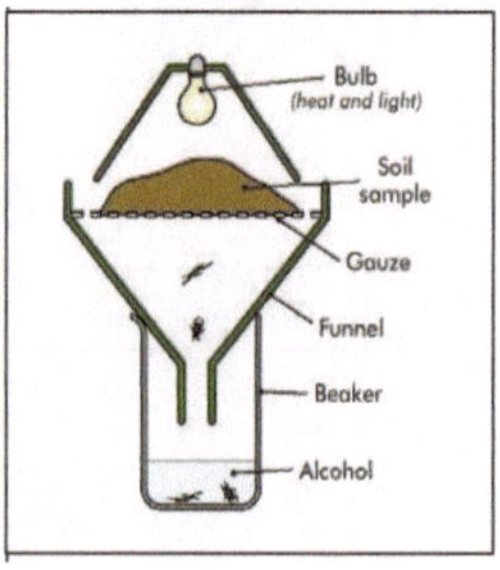

Berlese Funnel

Water Trap

Yellow Sticky Trap

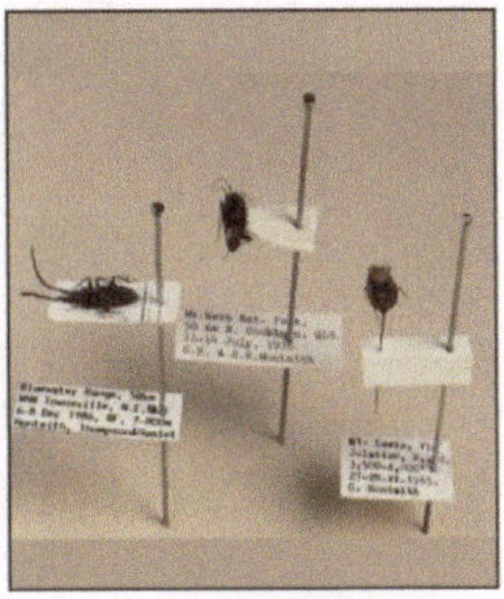

Carding

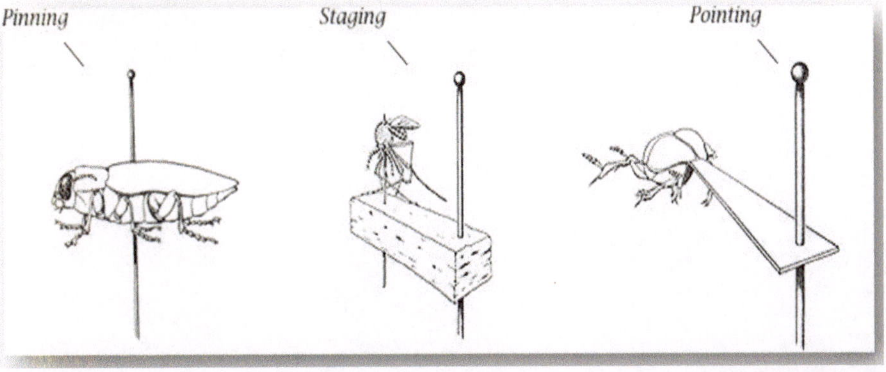

2

To Study the External Features of Grasshopper

The grasshoppers are widely distributed throughout the country and may be seen in abundance during monsoon season. For the generalized morphological description, this insect has been considered as the most suitable representative of class insects because its structural details are not much variable. Apart from this, being larger, it can be studied easily. The generalized insect body is divided into 3 distinct body regions: a head, a thorax and an abdomen. Grouping of body segments into distinct regions is known as **Tagmosis** and the body regions are called **Tagmata**.

The Head

This is an anterior part of the body formed by the fusion of six segments viz., ocellary, antennal, intercalary, mandibular, maxillary and labial. All these segments are closely amalgamated to form a hard case or head capsule, the cranium that bears the antennae, eyes and mouthparts. The head is attached to the thorax using a flexible membranous neck (cervix) that allows its movement. The head capsule is sclerotized and the head capsule excluding appendages formed by the fusion of several sclerites is known as **cranium**.

Sclerites of Head

1. **Vertex**: Summit of the head between compound eyes.
2. **Frons**: Facial area below the vertex and above the clypeus.
3. **Clypeus**: Cranial area below the fronts to which labrum is attached.
4. **Gena**: Lateral cranial area behind the compound eyes.
5. **Occiput:** Cranial area between occipital and post occipital suture.

Sutures of Head: The linear invaginations of the exoskeleton between two sclerites are called as suture (sometimes referred as sulcus).

1. **Epicranial suture/ ecdysial line**: Inverted `Y' shaped suture found medially on the top of the head, with a median suture (coronal suture) and lateral sutures (frontal suture).

2. **Epistomal suture/ Fronto clypeal suture**: Found between frons and clypeus. (epi –above; stoma- mouth parts)
3. **Clypeo-labral suture**: Found between clypeus and labrum (upper lip).
4. **Postoccipital suture**: Groove bordering occipital foramen. Line indicating the fusion of maxillary and labial segment.

The Thorax: It is a body region situated between the head and abdomen. The insect thorax is composed of three segments: an anterior prothorax, a middle mesothorax, and a posterior metathorax. Each segment bears a pair of legs. The last two segments often called pterothorax may bear wings. Meso and metathorax which bear wings are called **Pterothorax**. Thoracic segments are made up of three sclerites namely, dorsal body plate **tergum or nota**, ventral body plate **sternum** and lateral plate **pleuron**

Functions of thorax: Site of locomotion.

Abdomen: Abdominal segments are telescopic, highly flexible and are interconnected by a membrane called **conjunctiva**. Each abdominal segment is made up of only two sclerites namely dorsal body plate (tergum) and ventral body plate (sternum). In grass hopper eight pairs of spiracles are present in the first eight segments, in addition to a pair of tympanum in the first segment. Eight and ninth abdominal segments bears the female genital structure and ninth segment bears male genital structure. Abdominal appendages in adult insects are genital organs and cerci.

Function: Site of metabolism and reproduction.

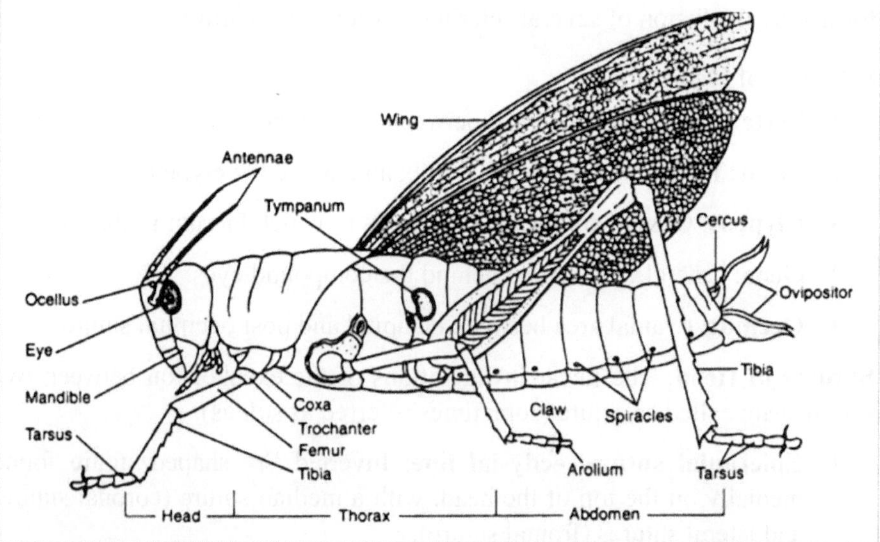

3

To Study the Types of Insect Antennae

The collected insect samples can be inspected in the laboratory after detaching the antenna and putting them under the microscope. They can also be studied through the permanent slides of different types of antennae with the help of microscope.

Antennae are mobile sensory segmented appendages of the head. They articulate with head in front or between the eyes and arise from antennal socket. The size and shape of antennae varies in different insects. They are used for sensory perception which includes motion and orientation, odor, sound, humidity, and a variety of chemical cues. The Sensilla on the antenna acts as tactile, olfaction, carbon dioxide, temperature, wind, humidity, and sound receptors.

Structure of Antenna: Antennae consist of three parts

1. **Scape-** It is the first basal segment of the antenna by which the antennae are attached to the head. It is often distinctly larger than the other succeeding joints. It articulates with the antennal ridge.
2. **Pedicel-** The joint immediately followed the shape of is pedicel. It is usually small and contains a special sensory structure known as **Johnston's organ**, which is absent in Diplura, Collembola.

Flagellum- It is also known as **clavola**, and is the remaining part of the antenna. Flagellum segments (flagellomeres) increase in number in certain insects. It is modified

a) According to the surroundings and habits of the insects.

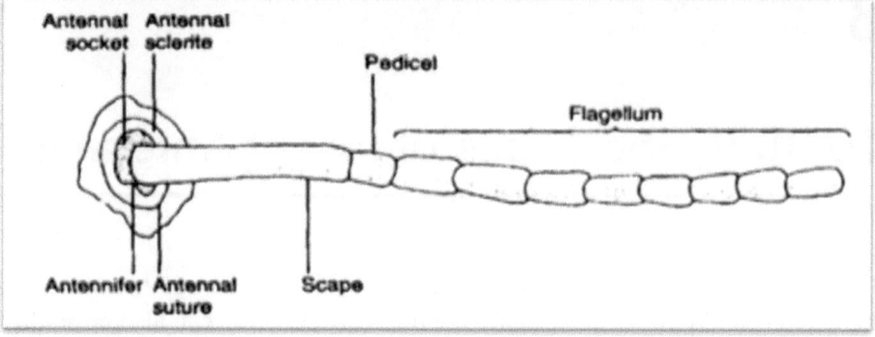

Types of antennae

1. **Setaceous**: (Bristle-like) Size of the segments decreases from base to apex. e.g. Leafhopper, Dragonfly, Damselfly.
2. **Filiform**: (Thread-like) Segments are usually cylindrical. The thickness of the segments remains the same throughout. e.g. Grasshopper.
3. **Moniliform**: (Beaded) Segments are either globular or spherical with prominent constriction in between e.g. Termite.
4. **Serrate**: (Saw-like) Segments have short triangular projections on one side. e.g. Longicorn battle
5. **Pectinate**: (Comb-like) Segments with long slender processes on one side e.g. Sawfly
6. **Bipectinate**: (Double comb-like) Segments with long slender lateral processes on both sides e.g. Silkworm moth
7. **Clavate**: (Clubbed) Antenna enlarges gradually towards the tip. e.g. Blister beetle
8. **Capitate**: (Knobbed) Terminal segments become enlarged suddenly e.g. butterfly
9. **Lamellate**: (Plate-like) Antennal tip is expanded laterally on one side to form flat plates e.g. lamellicorn beetle
10. **Aristate**: The terminal segment is enlarged. It bears a conspicuous dorsal bristle called arista e.g. House fly
11. **Stylate**: Terminal segment bears a style-like process eg. horse fly, rubber fly.
12. **Plumose**: (Feathery) Segments with long whorls of hairs e.g. male mosquito

13. **Pilose**: (Hairy) Antenna is less feathery with few hairs at the junction of flagellomeres. e.g. Female mosquito.

14. **Geniculate:** (Elbowed) Scape is long remaining segments are small and are arranged at an angle to the first resembling an elbow joint. e.g. Ant, weevil and honey bee.

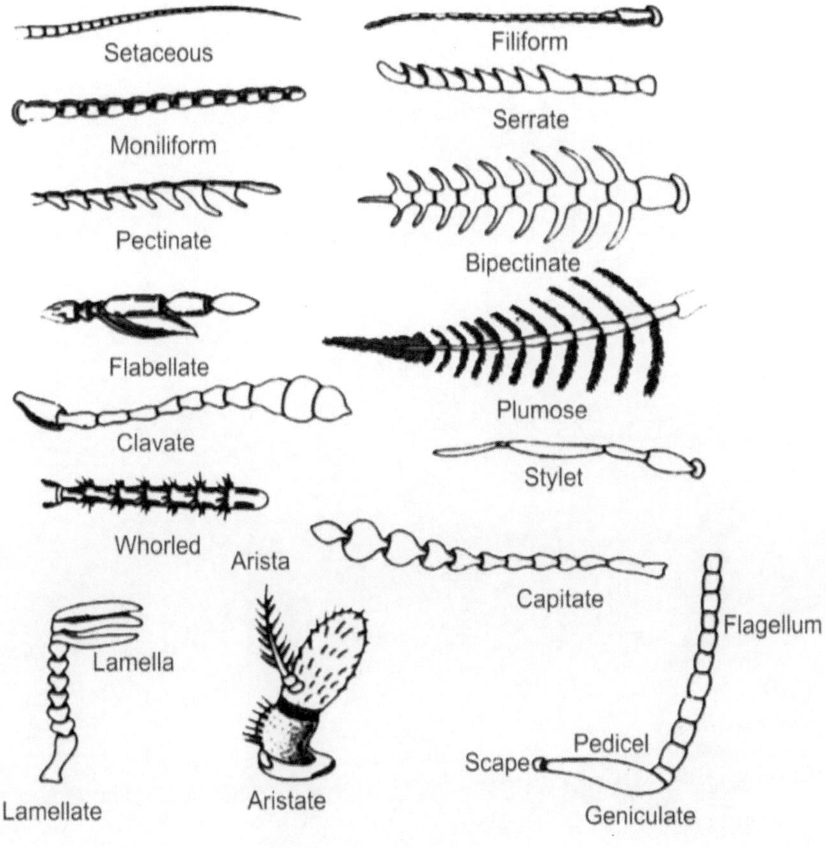

4

To Study the Different Types of Mouthparts and Their Modifications

Mouthparts of insects vary among insects of different groups depending upon their feedinghabits. They are mainly of two types viz., Mandibulate (feeding mainly on solid food) andhaustellate (feeding mainly on liquid food). Insect mouthparts have become modified in variousgroups to perform the ingestion of different types of food and by different methods. Indeed themodifications in the mouthparts to ingest almost all kinds of the food material, are one of thefactors for the success of the group.

1. Biting and chewing type: e.g. Cockroach & grasshopper.

It is the primitive type of mouth part and consists of the following parts.

i. **Labrum- (Upper lip):** It is flap like, bilobed and attached to the clypeus by an articular membrane. It is movable. It covers the mouth cavity from above. It helps to pull the food into the mouth. It holds the food in position so that mandibles can act on it. It forms the roof ofthe pre oral food cavity.

ii. **Labrum-epipharynx:** Inner surface of the labrum is referred to as epipharynx. It is frequently membranous and continuous with the dorsal wall of pharynx. It is an organ of taste.

iii. **Mandibles:** There is a pair of mandibles. They are the first pair of jaws. They are also called as primary jaws or true jaws. Mandibles articulate with the cranium at two points. They are heavily sclerotised. They are toothed on their inner border. There are two types of teeth. Distalare sharply pointed and are called incisor or cutting teeth and proximal teeth are called molar or grinding teeth. They act transversely to bite and grind the food into small fragments.

iv. **Maxillae:** They are paired and more complicated than mandibles. They are called secondary jaws or accessory jaws. At proximal end the first sclerite cardo joins the maxilla to head. The second sclerite is called stipes which articulates with cardo. Stipes carries a lateral sclerite called palpifer which bears a five segmented antenna like maxillary palp. On the distal end of the stipes, there are two lobes. The outer lobe is called galea and inner lobe is lacinia which is toothed. Maxille direct the food into the mouth. They hold the food in place when the mandibles are in action. They act as auxillary jaws and assist in mastication of food. Sense organs connected with the perception of touch, smell and taste are abundantly found in palpi.

v. **Hypopharynx:** It is a tongue like organ. It is located centrally in the preoral cavity. Salivary gland duct opens through it.

vi. **Labium- (lower lip):** It is a composite structure formed by the fusion of two primitive segmented appendages. It bounds the mouth cavity from below or behind. It forms the base of the pre oral cavity. It consists of three median sclerites viz., submentum(large basal sclerite), mentum (middle sclerite) and prementum (apical sclerite). On the lateral side of the prementum there are two small lateral sclerites called palpiger bearing three segmented labial palpi. Distally prementum bears two pairs of lobes. The other pair of lobes is called para glossae and inner pair of lobes, glossae. Both pairs when fused are called ligula.

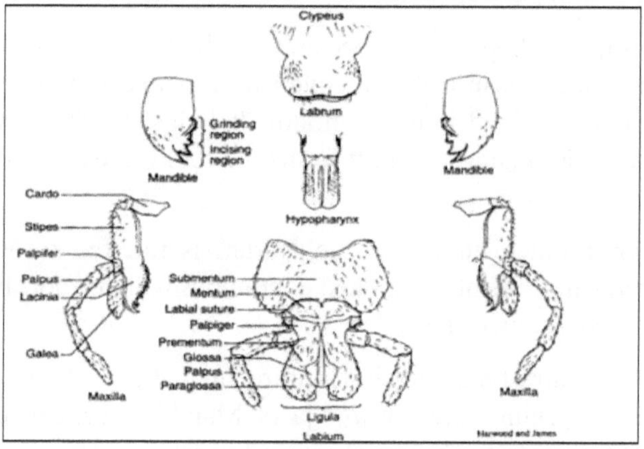

Figure: Biting and chewing type

2. Piercing and Sucking / Hemipteran /bug type e.g. Plant bugs

Labium projects downwards from the anterior part of the head like a beak. The beak is four-segmented and grooved throughout its entire length. At the base of the labium, there is a triangular flap-like structure called a labrum. Labium is neither involved in piercing nor sucking. It functions as a protective covering for the four styles (fascicle) found within the groove. Both mandibles and maxillae are modified into long slender sclerotized hair-like structures called stylets. They are lying close together and suited for piercing and sucking. The tips of the stylets may have minute teeth for piercing the plant tissue. The inner maxillary styles are doubly grooved on their inner faces. When these are closely opposed they form two canals viz., food canal and salivary canal through sap and saliva are conducted respectively. Saliva contains enzymes or toxins that can distort plant cell walls to permit the stylets to penetrate down and reach the phloem for suking the sap. Both palps are absent.

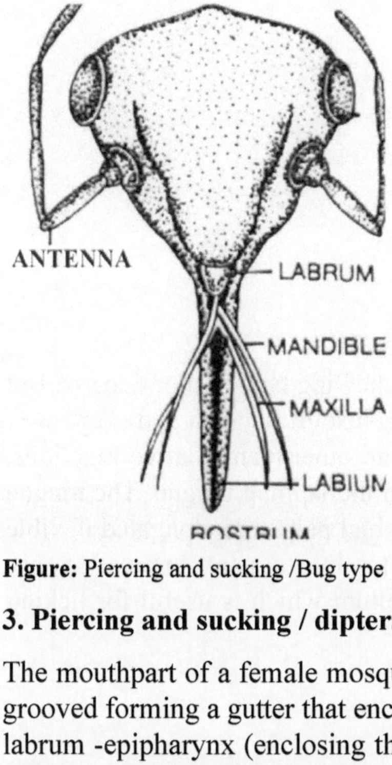

Figure: Piercing and sucking /Bug type

3. Piercing and sucking / dipteran /mosquito type: e.g. Female mosquito

The mouthpart of a female mosquito consists of an elongate labium which is grooved forming a gutter that encloses six styles. The stylets are composed of labrum -epipharynx (enclosing the food canal), the hypopharynx (containing the salivary canal), two maxillae and two mandibles. Both the ends of maxillary stylets and mandibular stylets are saw-like and suited to piercing flesh. The stylets are inserted into the host's skin by a strong downward and forward thrust of the body. Both mandibles and maxillae are reduced in males and they feed on plant nectar and juices of decaying fruits. A female pierces the skin of human beings into which it injects saliva containing an anticoagulant (to keep the blood flowing without clotting) and an anesthetic (to keep the victim unaware of the bite) and sucks up the blood. Labium does not pierce but folds up or back as stylets pierce. Maxillary palpi are present.

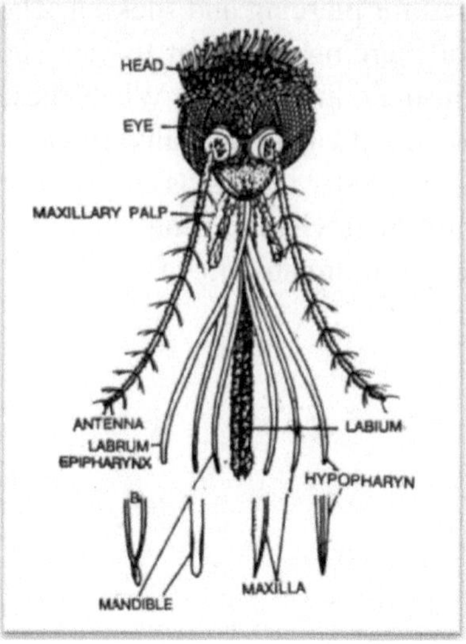

Figure: Piercing and sucking /Bug type

4. Chewing and lapping type: e.g. honey bee

Labrums and mandibles are as in biting and chewing type of mouth parts. But mandibles are blunt and not toothed. They are useful to crush and shape wax for comb building; ingesting pollen grains and other manipulative functions. Maxillofacial structures are modified to form the lapping tongue. The tongue unit consists of two galea of maxillae, two labial palpi and elongated flexible hairy glossa of labium. The glossa terminates into a small circular spoon-shaped lobe called spoon or bouton or flabellum which is useful for licking the nectar.

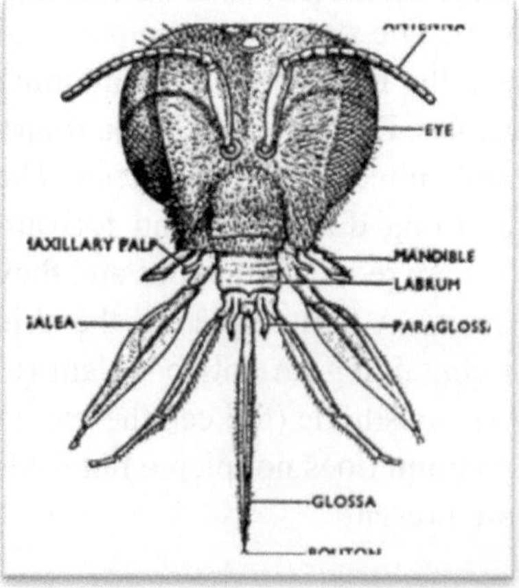

Figure: Chewing and lapping type/ Honeybee

5. Rasping and sucking: e.g. Thrips

The mouth cone consists of the labrum, labium, and maxillae. There are three stylets derived from two maxillae and left mandible. The right mandible is absent. Stylets are useful to lacerate the plant tissue and the oozing sap is sucked up by the mouth cone. Both maxillary palpi and labial palpi are present.

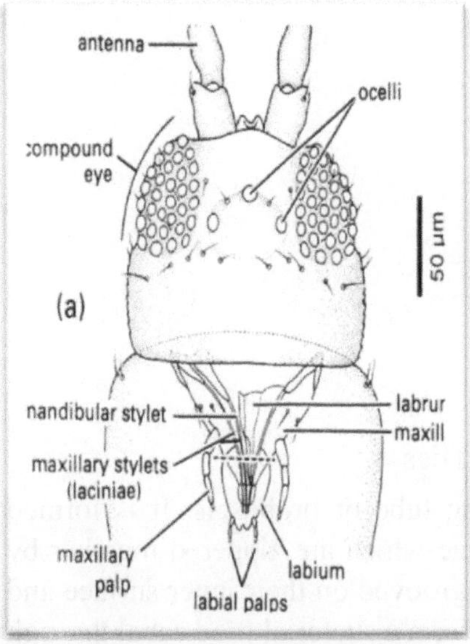

Figure: Rasping and sucking: e.g. Thrips

6. Sponging type: e.g. House fly

The proboscis is fleshy, elbowed, retractile and projects downwards from the head. The proboscis can be differentiated into basal rostrum and distal haustellum. The proboscis consists of labium which is grooved on its anterior surface. Within this groove lie the labrum-epiphraynx (enclosing the food canal) and slender hypopharynx (containing the salivary canal).Mandibles are absent. Maxillae are represented by single-segmented maxillary palpi. The end of the proboscis is enlarged, sponge-like and two-lobed which acts as suction pads. They are called oral discs or labella. The surfaces of labella are transvered by capillary canals called pseudo tracheae which collect the liquid food and convey it to the canal. Labella function as sponging organs and are capable of taking exposed fluids. These insects often spit enzyme-containing saliva onto solid foods to liquify them.

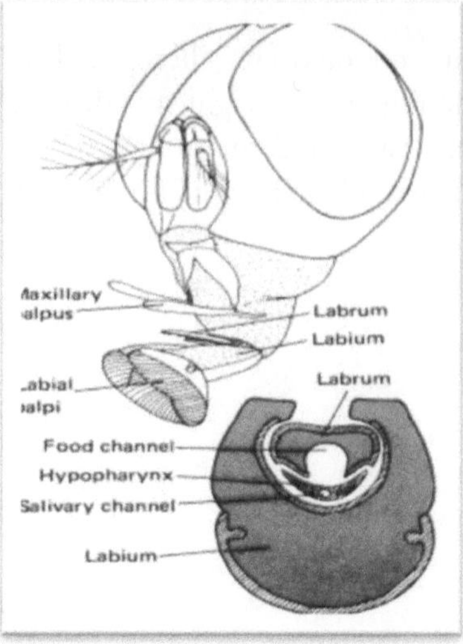

Figure: Sponging type/ Housefly

7. Siphoning type: e.g. Moths and butterflies

Mouthparts consist of an elongate sucking tube or proboscis. It is formed by two greatly elongated galeae of maxillae which are zippered together by interlocking spines and hooks. Galeae are grooved on their inner surface and when they are fitting together closely they form a suctorial food canal through which the nectar is sucked up. The proboscis is coiled up like watch spring and kept beneath the head when it is not in use.

By pumping of blood into galeae, the proboscis is extended. The other mouth parts are reduced or absent except the labial palpi and smaller maxillary palpi.

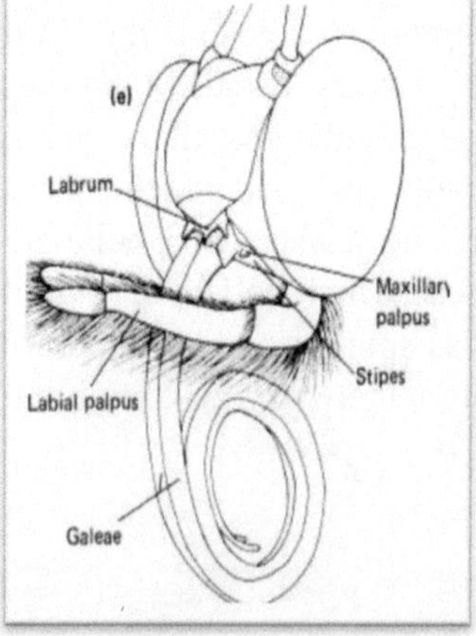

Figure: Siphoning type/ Moths and butterflies

5

To Study the Different Legs and Their Modifications

The typical thoracic leg consists of six parts, basal coxa that articulates with the thorax in the pleural region, small trochanter, femur, tibia, segmented tarsus, and pre-tarsus. The coxa is often divided into two parts, the posterior and the anterior (usually the larger part) being called the meron. The trochanter articulates with the coxa but usually forms an immovable attachment with the femur. The femur and tibia are typically the longest leg segments. The tarsus, which is derived from a single segment, - is usually subdivided into individual tarsomeres. The pretarsus may consist of a single claw, but it is usually composed of a pair of moveable claws and one or more pads or bristles. Legs are usually looked upon as the principal organs of terrestrial locomotion. They have undergone many modifications and have been adapted to a wide variety of functions including swimming, prey capture, pollen collection and digging

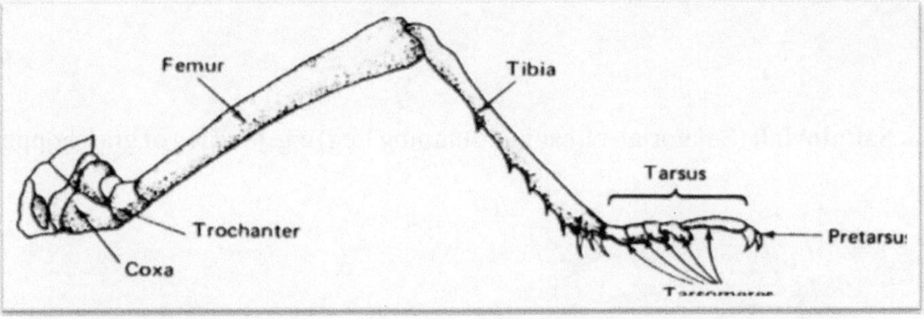

1. Ambulatorial (Ambulate - to walk; Walking leg) e.g. Fore leg and middle leg of grasshopper. Femur and tibia are long. Legs are suited for walking

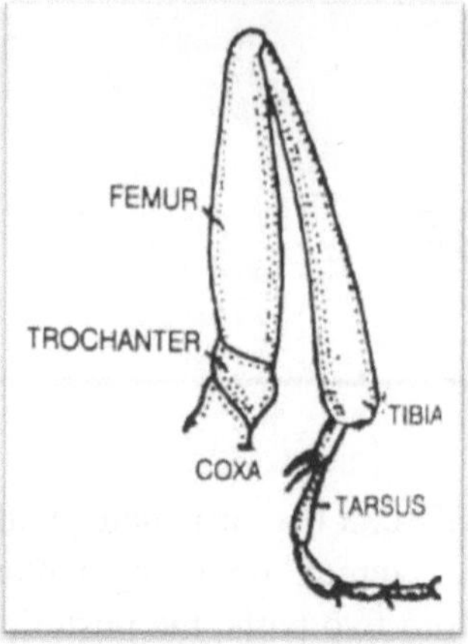

2. Cursorial: (Cursorial = adapted for running: Running leg) e. g. All the three pairs of legs of the cockroach. Legs are suited for running. The femur is not swollen.

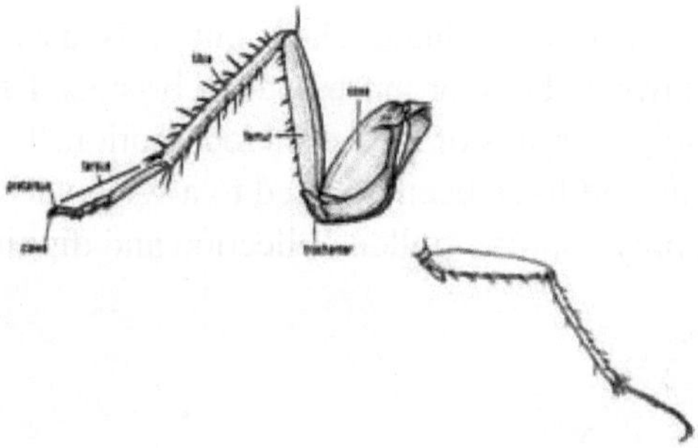

3. Saltatorial: (Salatorial = Leaping: Jumping Leg) e.g. hind leg of grasshopper.

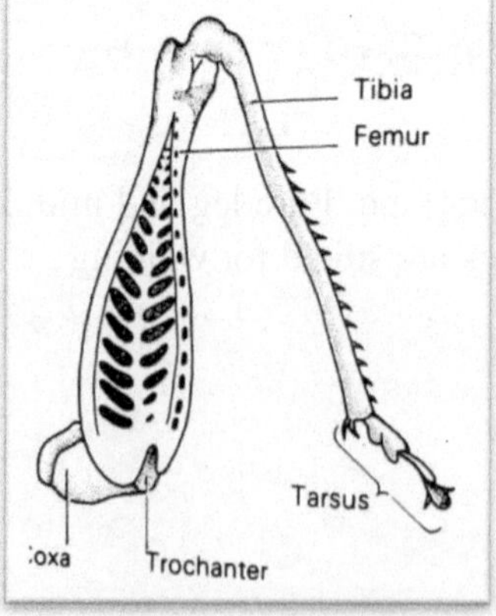

4. Fossorial: (Forrorial = Digging; Burrowing leg) e.g. Fore legs of mole cricket.

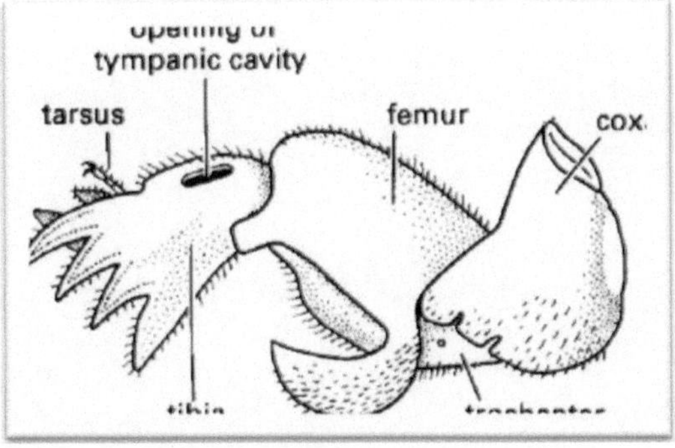

5. Natatorial: (Natatorial = pertaining to swimming; Swimming leg) e.g. hing legs of water bug and water beetle.

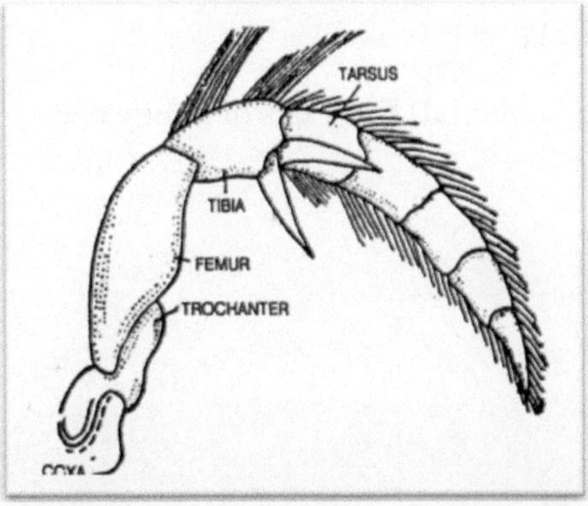

6. Raptorial: (Raptorial=predatory; Grasping leg) e.g. Forelegs of preying mantids.

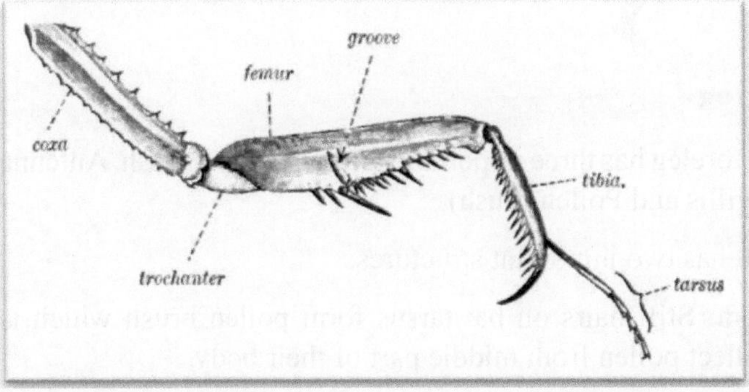

7. Scansorial: (Scansorial = Climbing; climbing or clinging leg) e.g. all the three pairs of legs of head louse.

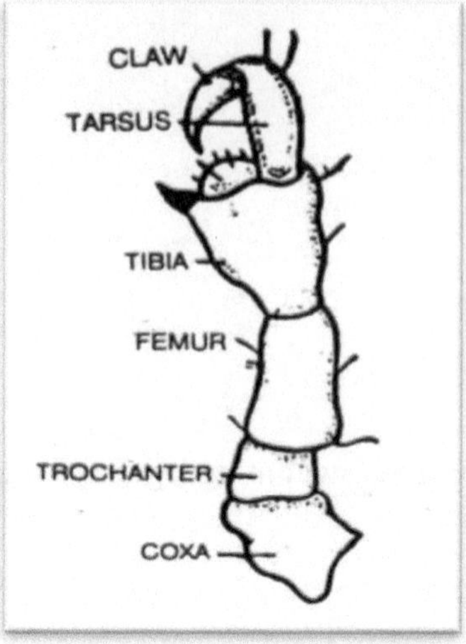

8. Foragial leg: (Forage = to collect food material) e.g. Legs of honey bee.

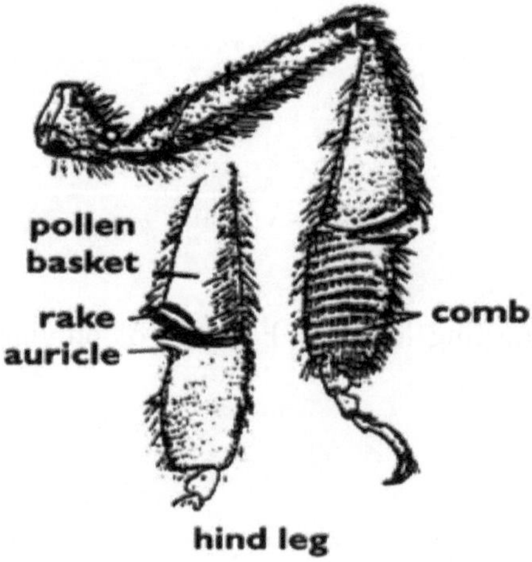

i. **Forelegs**: The foreleg has three important structures (Eye brush, Antenna cleaner or strigillis and Pollen brush)

ii. **Middle legs**: It has two important structures.

(a) **Pollen brush**: Stiff hairs on basitarsus form pollen brush which is useful to collect pollen from middle part of their body.

(b) **Tibial spar**: At the distal end of the tibia, a movable spur is present which is useful to loosen the pellets of pollen from the pollen basket of hind legs and to clean wings and spiracles.

iii. **Hind legs**: It has three important structures viz., pollen basket, pollen packer and pollen comb.

(a) **Pollen basket**: It is also called corbicula. The outer surface of the hind tibia contains a shallow cavity. The edges of the cavity are fringed with long hairs. The pollen basket enables the bee tocarry a larger load of pollen and propolis from the field to the hive.

(b) **Pollen packer**: It is also called pollen press. It consists of pecten and auricle. Pecten is a row of stout bristles at the distal end of tibia. Auricle is a small plate

9. Climbing or Sticking leg: e.g. all the three pairs of legs of house fly.

10. Clasping leg: e.g. Forelegs of male water beetle.

Prolegs: Caterpillars 'possess three pairs of thoracic legs (true-legs) and five pairs of abdominal legs (pro-legs) on 3rd, 4th, 5th, 6th, and last abdominal segments. In some, semi-loopers larvae, prologs on 3rd and 4th abdominal segments are absent, and hence while movement, it looks like semi loop, in some, loppers, prologs are present only on 6th and last abdominal segments, and hence while movement, it looks like a loop.

- Thoracic legs are also called the true legs, which are typically jointed and sclerotized.
- The abdominal legs are called prolegs. These are unjointed, short, and fleshy with a flat surface at the bottom called planta.
- Several hooks-like structures called crochets are seen arranged in circular or semi-cuticular form on the surface of the plants.
- In sawflies of Hymenoptera, the larvae have 3 pairs of true legs in the thorax and 6 or >6 pairs of prologs in the abdomen. This is the unique feature of sawfly larvae, but these prologs do not bear crochets, unlike lepidopteran larvae.

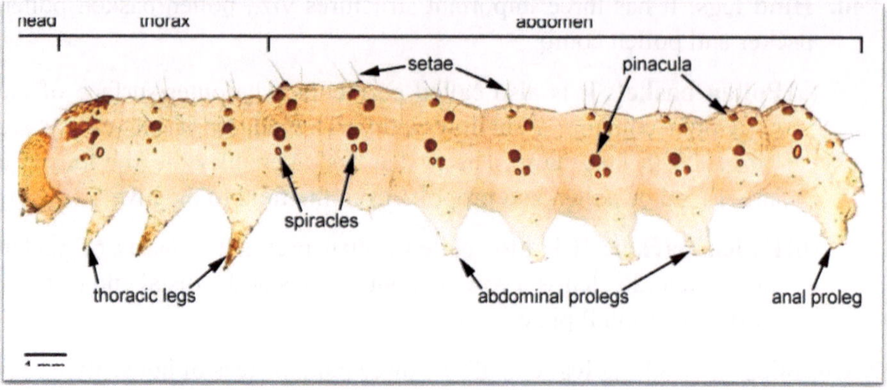
setae
pinacula
spiracles
thoracic legs
abdominal prolegs
anal proleg

6

To Study the Wing Venation, Types of Wings and Wing Coupling Apparatus

The complete system of veins of a wing is termed venation or neuration. Generally in all the insects, there is some similarity in wing venation and therefore, it is presumed that all types of wing venation have developed from the common base or the same ancestor. Using an extensive study of wing venation in different groups of insects, **Comstock and Needham** constructed a hypothetical type of wing venation from which all other types have presumably been derived. According to them the primitive wing venation has developed from two tracheae which are situated on the anterior and the posterior basal margins of wings and their branches are spread all over the wing. Each main trachea gives rise to three principal veins, thereby forming 6 principal veins namely costa, radius, medius, cubitus, penultimate and ultimate. Each principal vein gives rise to a sub-vein near its base.

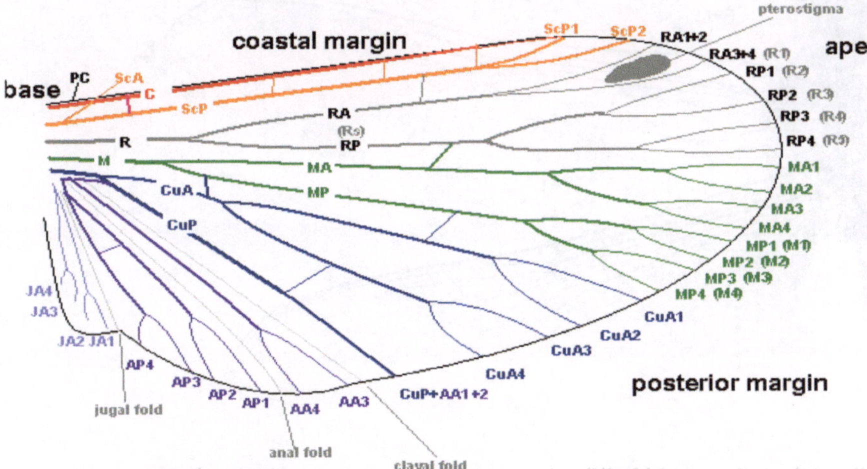

Wing Venation

Among the invertebrate animals, only insects possess wings. Wings are present only in the adult stage. Number of wings varies from two pairs to none. Certain primitive insects like silverfish and springtails have no wings (apterous). Ectoparasites like head louse, poultry louse and flea are secondarily wingless. Wings are deciduous in ants and termites. There is only one pair of wings in the true flies. Normally two pairs of wings are present in insects and they are borne on pterothoracic segments viz., mesothorax and metathorax. Wings are moved by thoracic flight muscles attached to their bases. The wing is a flattened double-layered expansion of the body wall with a dorsal and ventral lamina having the same structure as the integument. Both dorsal and ventral laminae grow, meet and fuse except along certain lines. Thus a series of tracheae, nerves and blood. Wing is nourished by blood circulating through veins. Later the walls of these channels become thickened to form veins or nerves. The arrangement of veins on the wings is called venation which is extensively used in insect classification. The principal longitudinal veins arranged in order from the anterior margin are costa (C), sub costa (Sc), radius ®, median (M), cubitus (Cu) and anal veins (A). Small veins often found interconnecting the longitudinal veins are called cross veins. Due to the presence of longitudinal veins and cross veins, the wing surface gets divided into several enclosed spaces termed cells. In insects like dragonfly and damselfly, there is an opaque spot near the coastal margin of the wing called pterostigma.

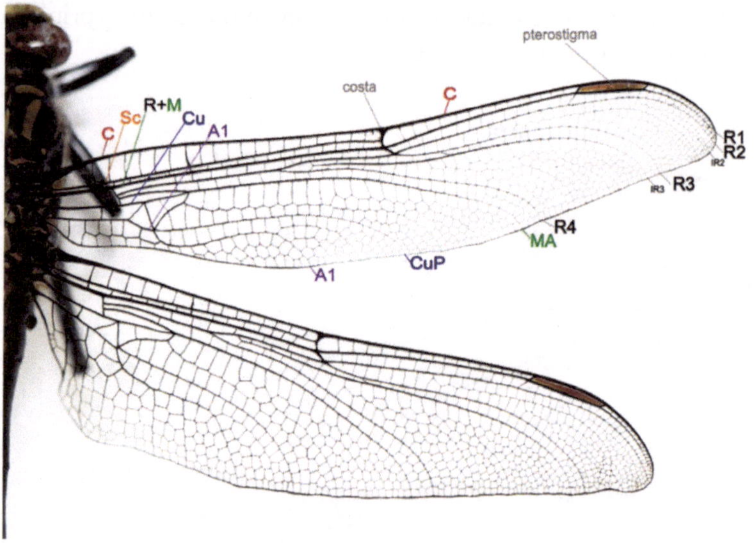

Wing Adaptations and Modifications

1. **Membranous Wings**

They are thin & more or less transparent, but some are darkened. These wings are with highly developed venation. They are useful for flight. Examples:

Hind wings of Grasshopper, Dragonflies and damselflies (order: Odonata), lacewings (order: Neuroptera), flies (order: Diptera), bees and wasps (order Hymenoptera), termites (order Isoptera)

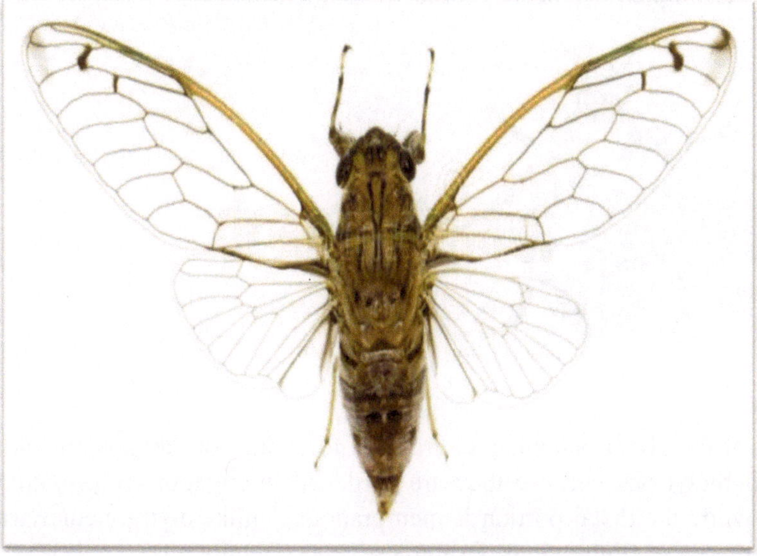

2. **Halters**

Halteres are an extreme modification among the order Diptera (true flies), in which the hind wings are reduced to mere nubs. Used for balance and direction during flight (gyroscopic stabilizers). It is divided into three regions-scabellum, pedicel and capitalum. Examples: All flies (Diptera).

3. **Elytra**

Elytra (singular elytron) are the hardened, heavily sclerotized forewings of beetles (horny sheet without venation). Modified to protect the hind wings when at rest. Examples: All beetles (order Coleoptera).

4. **Hemelytra**

A variation of the elytra is the hemelytra. The forewings of Hemipterans are said to be hemelytrous because they are hardened throughout the proximal two-thirds, while the distal portion is membranous. Unlike elytra, hemelytra function primarily as flight wings. Examples: Bugs (Hemiptera).

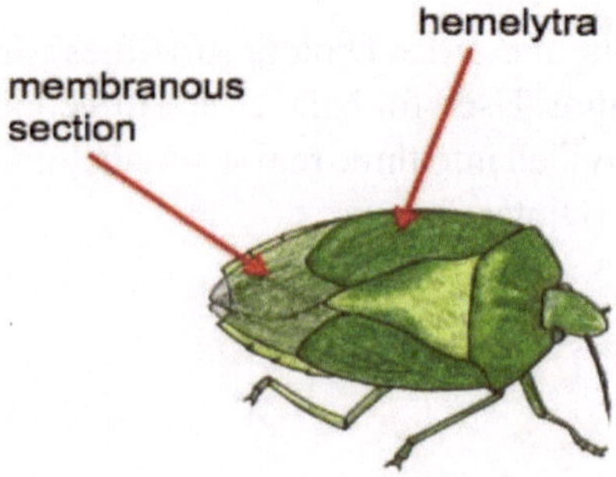

5. **Tegmina**

Tegmina (singular tegmen) are the leathery forewings of insects in the orders Orthoptera, Dictyoptera. Like the elytra on beetles the tegmina help protect the delicate hindwings, and also used forflight sometimes. Examples: Grasshoppers, crickets and katydids (Orthoptera), Cockroaches, Mantids (Dictyoptera).

6. Scaly Wings

Scaly wings-thin and membranous front andhind wings covered oversurface with flattened unicellular setae(scales). The scales make the wings colorful and usedfor taxonomic studies. They are useful forflight. Examples: Butterflies, moths and skippers (order Lepidoptera), caddisflies (order: Trichoptera).

7. Fringed Wings

Fringed wings-slender front and hind wings with long fringes of marginal hairs, giving a feather-like appearance. The wings are highly reduced with reduced venation. They are useful for flight. Example: Thrips (Thysanoptera).

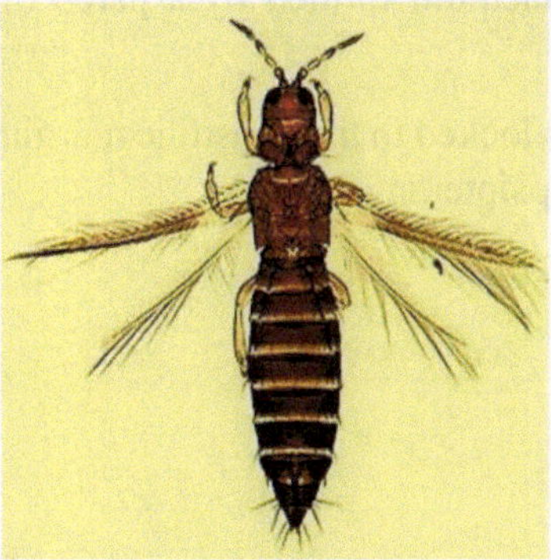

8. Clefted Wings (Fissured wings)

Front wing is longitudinally divided forming a fork-likestructure. The hind wing is dividedtwice, forming two forks with three arms. All forks possess smallmarginal hairs. They are useful for flight. Example: Both wings of Plume Moth.

Wing Coupling

Higher pterygotes have attained virtual dipterism by co ordinate wing movements. Such insects have devices for hooking fore and hind wings together so both the pairs move synchronously. By coupling the wings the insects become functionally two winged.

Types of Wing Coupling

1. **Hamulate:** A row of small hooks is present on the costal margin of the hind wing which is known as hamuli. These engage the folded posterior edge of fore wing. Eg: Bees.
2. **Amplexiform:** It is the simplest form of wing coupling. A linking structure is absent. Coupling is achieved by broad overlapping of adjacent margins. Eg: Butterflies.
3. **Frenate:** There are two subtypes. Eg: Fruit sucking moth.

   (a) Male frenate: Hind wing bears near the base of the costal margin a stout bristle called frenulum which is normally held by a curved process, retinaculum arising from the sub-costal vein found on the under surface of the forewing.

   (b) Female frenate: Hindwing bears near the base of the costal margin a group of stout bristle (frenulum) that lies beneath the extended forewing and engages there in a retinaculum formed by a patch of hairs near cubitus.

Jugate- Jugam of the forewing are lobe and it is locked to the coastline margin of the forewing. Example: Hepialid moths, strepsipteran.

7

To Study the Types of Insect Larvae and Pupae

Metamorphosis and immature stages

The change in growth and development (form) of an insect during its life cycle from birth to maturity is called metamorphosis. There are four basic types of metamorphosis in insects.

Ametabola: (No metamorphosis) e.g. Silver fish.

These insects have only 3 stages in their life cycle namely egg, young ones and adults. It is most primitive type metamorphosis. The hatching insects resemble the adult in all respects except for the size and called as juveniles. Moulting continues throughout the life.

Hemi-metabola: (Incomplete metamorphosis) e.g. Dragonfly, damselfly and may fly.

These insects also have 3 stages in their life namely egg, young ones and adults. The young ones are aquatic and are called **naiads**. They are different from adults in habits and habitat. They breathe by means of tracheal gills. In dragonfly naiad, the lower lip (labium) is called **mask** which is hinged and provided with hooks for capturing prey. After final moult, the insects have fully developed wings suited for aerial life.

Pauro-metabola: (Gradual metamorphosis) e.g. Cockroach, grasshopper, bugs.

The young ones are called nymphs. They are terrestrial and resembles the adults in general body form except the wing and external genitalia. Their compound eyes and mouthparts are similar to that of adults. Both nymphs and adults share the same habitat. Wing buds externally appear in later instars. The genitalia development is gradual. Later instars nymphs closely resemble the adult with successive moults.

Holo-metabola: (Complete metamorphosis) e.g. Butterflies, moths, fly and bees.

These insects have 4 stages namely egg, larva, pupa and adult. Majority of the insects undergo complete metamorphosis. Larvae of butterfly is called caterpillar. Larva differs greatly in form from adult. Compound eyes are absent in larva. Lateral ocelli or stemmata are the visual organs. Their mouth parts and food habits differ from adults. Wing development is internal. When the larval growth completed it transforms into pupa. It is the resting and non feeding stage in which the larval tissues disintegrate and adult organs are built up.

Larvae: There are three main types of insects larvae namely oligopod, polypod and apodous.

Oligopod: Thoracic legs are well developed. Abdominal legs are absent. There are subtypes:

a) **Campodeiform:** They are so called from their resemblance to the dipluran genus campodee. Body is elongate, depressed dorso ventrally and well sclerotised. Head is prognathous. Thoracic legs are long. A pair of abdominal cerci or caudal processes is usually present. Larvae are generally predators and are very active. Eg: grub of ant lion or grub of lady bird beetle.

b) **Scarabaeiform:** Body is "C'shaped, stout and subcylindrical. Head is well developed. Thoracic legs are short. Caudal processes are absent. Larva is sluggish, burrowing into wood or soil. Eg: grub of rhinocerous beetle.

Polypod or Eruciform

The body consists of an elongated trunk with large sclerotized head capsule. The head bears a pair of powerful mandibles which tear up vegetation. Two groups of single-lensed eyes found on either side of the head constitute the visual organs. The antenna is short. Three pairs of thoracic legs and up to five pairs of unjointed abdominal legs or prologs are present. Thoracic legs are segmented and they end in claws which are used for holding typically bears rows or circles of short hooked spines or crochets which are useful in clinging to the exposed surface of vegetation and walking. Abdominal segments three to six and ten typically bear prologs. Eg: caterpillar (larvae of moths and butterflies).

a. **Hairy caterpillar:** The body hairs may be dense, sparse or arranged in tufts. Hairs may cause irritation, when touched. Eg: Red hairy caterpillar.

b. **Slug caterpillar:** Larva is thick, short, stout and fleshy. Larval head is small and retractile. Thoracic legs are minute. Abdominal legs are absent. Abdominal segmentation is indistinct. Larva has poisonous spines called scoli distributed all over the body. Such larva is also called platyform larva.

c. **Semilooper:** Either three or four pairs of prologs are present. Prologs are either wanting or rudimentaly in either third or third and fourth abdominal segments. Eg: Castor semilooper.

d. **Looper:** They are also called measuring worm or earth measurer or inch worm. In this type only two pairs of prologs are present in sixth and tenth abdominal segments. Eg: Daincha looper.

Apodous

They are larvae without appendages for locomotion. Based on the degree of development and sclerotization of the head capsule, there are three subtypes.

Eucephalous: larva with a well-developed head capsule with functional mandibles, maxillae stemmata and antennae. Mandibles act transversely. Eg: Wriggler (larva of mosquito).

Hemicephalous: Head capsule is reduced and can be with drawn into thorax. Mandibles act vertically. Eg: Larva of horse fly and robberfly.

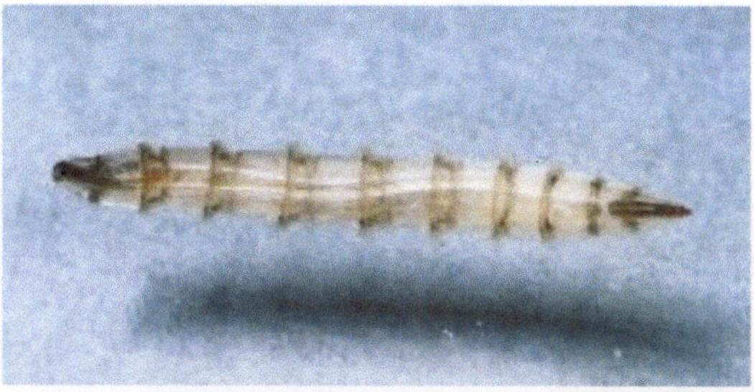

Acephalous: Head capsule is absent. Mouth parts consist of a pair of protrusible curved mouth hooks and associated internal sclerites. They are also called vermiform larvae.

Eg: maggot (larva of housefly).

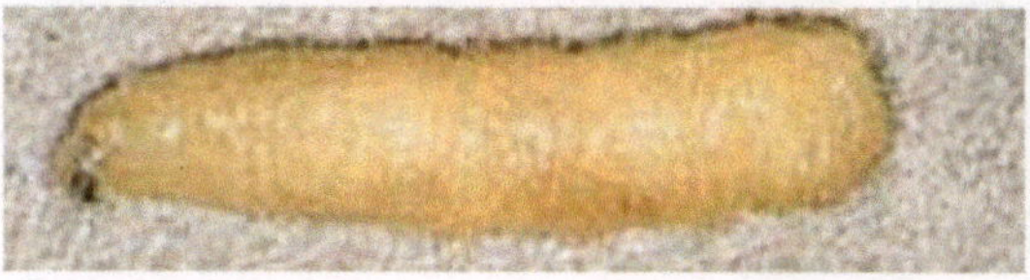

Pupae

It is the resting and inactive stage in all holometabolous insects. During this stage, the insect is incapable of feeding and is quiescent. During this transitional stage, the larval characters are destroyed and new adult characters are created. There are three main types of pupae.

I. Obtect: Various appendages of the pupa viz., antennae, legs and wings pads are glued tothe body by a secretion produced during the last larval moult. Exposed surface of the appendages are more heavily sclerotised than those adjacent to body. eg., moth pupa.

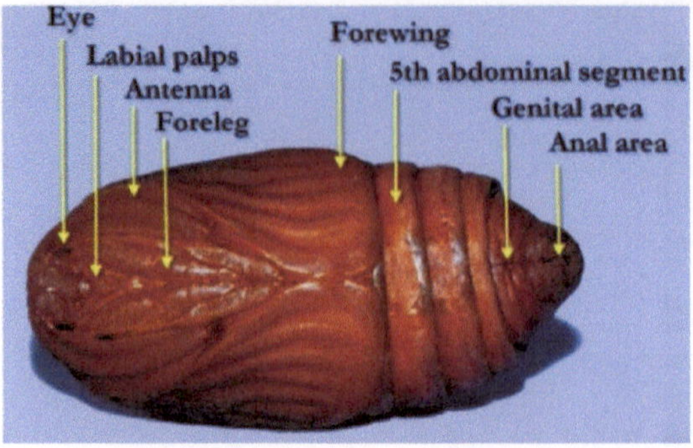

a. **Chrysalis:** It is the naked object pupa of the butterfly. It is angular and attractively colored. The pupa is attached to the substratum by hooks present at the terminal end of the abdomen called cremaster. The middle part of the chrysalis is attached to the substratum by two strong silken threads called girdle.

b. **Tumbler:** Pupa of a mosquito is called a tumbler. It is an object type of pupa. It is comma-shaped with rudimentary appendages. Breathing trumpets are present in the cephalic end and anal paddles are present at the end of the abdomen. The abdomen is capable of jerky movements which are produced by the anal paddles. The pupa is very active.

II. Exarate: Various appendages viz., antennae, legs and wing pads are not glued to the body. They are free. All oligopod larvae will turn into exarate pupae. The pupa is soft and pale. E.g., pupa of rhinoceros beetle.

III. Coarctate: The pupal case is barrel shaped, smooth with no apparent appendages. The last larval skin is changed into a case containing the exarate pupa. The hardened dark brown pupal case is called puparium. eg., Fly pupa.

Pupal Protection

In general pupal stage lacks mobility. Hence, it is the most vulnerable stage. To get protection against adverse conditions and natural enemies, the pupa is enclosed in a protective cover called cocoon. Based on the nature and materials used for the preparation of cocoons, there are several types:

S. No.	Types of cocoon	Materials used	Example
1.	Silken cocoon	silk	silkworm
2.	Earthern cocoon	Soil + saliva	Gram pod borer
3.	Hairy cocoon	Body hairs	Wolly bear
4.	Frassy cocoon	Frass + saliva	Coconut black headed caterpillar
5.	Fibrous cocoon	Fibres	Red palm weevil
6.	Puparium	Hardened last larval skin	House fly

8

To Study the Dissection of Digestive System in Insects (Grasshopper)

The best learning situation requires one specimen and set of tools per two students for studies. Students working in pairs have ample opportunity to fully participate in the dissection and to carefully examine the specimen. They are also able to share and discuss their observations during and after the dissection.

Requirements: Dissecting pan with wax pad, dissecting pins, forceps, hand lenses and scissors

Procedure

Place the specimen in the dissecting pan ventral side up. Use scissors to cut through the exoskeleton‘s ventral side from the head to the posterior end of the abdomen.

Pull the cut sides apart and observe the internal organs. Pin the sides of the insect to the dissecting pan. The Observation of the digestive system is as follows:

Digestion takes place in specialized cavities joined together to form a continuous canal. It is called the alimentary canal.

The alimentary canal is divided into three main portions: Foregut, Midgut, and Hindgut

Foregut: It consists of the mouth surrounded by the mouthparts. The mouth cavity is called the pharynx. It continues as the esophagus that is short, narrow, and thin-walled. The canal then enlarges into a crop which is also thin-walled. The crop opens into the short, muscular organ, the gizzard or the proventriculus. A pair of salivary glands lie outside and below the crop.

Each salivary gland is branched the secretions of all the branches pouring into a common duct. The two ducts, one of each side, open into the mouth cavity at the labium. The entire foregut is lined with chitin. In the gizzard, the chitin forms teeth and plates to facilitate the grinding of the food.

Midgut: The Midgut consists entirely of the stomach or ventriculus. At the junction of the gizzard and stomach are six pairs of gastric caecae. These are pouch-like structures arranged in a ring-like manner around the anterior end of the stomach. The anterior lobe of each pair of the caecae extends over the proventriculus and the posterior lobe extends over the ventriculus.

The caecae secrete digestive juices and pour them into the stomach. The midgut is not lined by chitin or cuticle but by a peritrophic membrane. This membrane protects the stomach wall from abrasions and is fully permeable to enzymes and digested food.

Hindgut

Hindgut is a coiled structure consisting of anterior ileum, middle colon and posterior rectum. The rectum opens to the exterior through the anus. The hindgut is lined with cuticle. At the junction of the stomach and ileum are attached numerous long tubules called the Malpighian tubules.

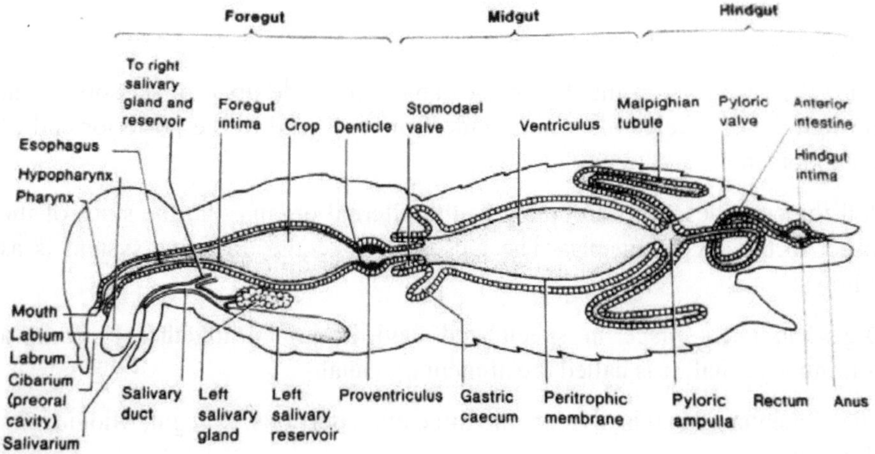

Diagram of Alimentary canal showing the major subdivisions in a generalized Grasshopper Insect

9

To Study the Male and Female Reproductive Systems in Insects

Requirements: Dissecting pan with wax pad, dissecting pins, forceps, hand lenses and scissors

Procedure

Place the specimen in the dissecting pan ventral side up. Use scissors to cut through the exoskeleton's ventral side from the head to the posterior end of the abdomen.

Pull the cut sides apart and observe the internal organs. Pin the sides of the insect to the dissecting pan. The observation of the reproductive systems of males and females are as follows:

Female reproductive system

The main functions of the female reproductive system are egg production and storage of male spermatozoa until the eggs are ready to be fertilized. The basic components of the female system are paired ovaries, which empty their mature oocytes (eggs) via the calyces (Calyx) into the lateral oviduct which unite to form the common (median) oviduct.

The gonopore (opening) of the common oviduct is usually concealed in an inflection of the body wall that typically forms a cavity, the genital chamber. This chamber serves as a copulatory pouch during mating and thus is often known as the bursa copulatrix. Its external opening is the vulva. In many insects the vulva is narrow and the genital chamber becomes an enclosed pouch or tube referred to as the Vagina.

Two types of ectodermal glands open into the genital chamber. The first is the spermatheca which stores spermatoza until they are needed for egg fertilization. The spermatheca is single and sac-like with a slender duct and often has a diverticulum that forms a tubular spermathecal gland. The gland or glandular cells within the storage part of the spermatheca provide nourishment to the contained spermatozoa.

The second type of ectodermal gland, known collectively as accessory glands, opens more posteriorly in the genital chamber.

Each ovary is composed of a cluster of egg or ovarian tubes, the ovarioles, each consisting of a terminal filament, a germarium (in which mitosis gives rise to primary oocytes), a vitellarium (in which oocytes grow by deposition of yolk in a process known as vitellogenesis) and a pedicel. An ovariole contains a series of developing oocytes each surrounded by a layer of follicle cells forming an epithelium (the oocyte with its epithelium is termed a follicle), the youngest oocytes occur near the apical germarium and the most mature near the pedicel.

The different types of ovariole are based on how the oocytes are nourished.

1. **Paniosticovariole**: Lacks specialized nutritive cells so that it contains only a string of follicles, with the oocytes obtaining nutrients from the hemolymph via the follicular epithelium. E.g. Cockroach.

Ovarioles of the other two contain trophocytes (nurse cells) that contribute to the nutrition of the developing oocytes.

1. **Telotrophicovariole (Acrotrophic)**: The trophocytes are confined to the germarium and remain connected to the oocytes by cytoplasmic strands as the oocytes move down the ovariole, e.g.bugs.
2. **Polytrophic ovariole**: Several trophocytes are connected to each oocyte and move down the ovariole with it, providing nutrients until depleted, thus individual oocytes alternate with groups of smaller trophocytes. E.g. moths andflies.

Accessory glands of the female reproductive tract are often called as colleterial or cement glands because their secretions surround and protect the eggs or cement them to the substrate. e.g. egg case production in mantis, ootheca formation in cockroaches, and Venom production in bees.

Structure of egg: Chorion, Vitelline membrane, Micropyle and Periplasm with yolk

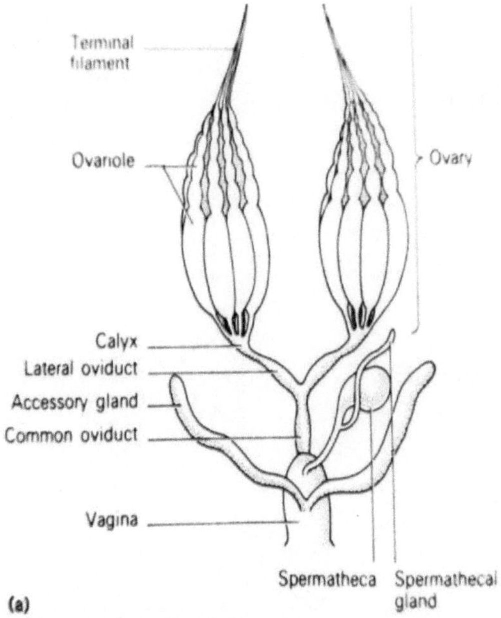

Female reproductive system

Male reproductive system

The main functions of the male reproductive system are the production and storage of spermatozoa and their transport in a viable state to the reproductive tract of the female. Morphologically, the male tract consists of paired testes, each containing a series of testicular tubes or follicles (in which spermatozoa are produced) that open separately into the mesodermally derived sperm duct or Vas deferens which expands posteriorly to form a sperm storage organ or seminal vesicle. Tubular paired accessory glands are formed as diverticula of the vasa deferentia. Sometimes the vasa deferentia themselves are glandular and fulfil the functions of accessory glands. The paired vasa deferentia unite where they lead into the ectodermally derived ejaculatory duct (the tube that transports the semen or the sperm to the gonopore).

Accessory glands are 1-3 pairs, either mesodermal or ectodermal in origin and associated with vasa deferentia or ejaculatory duct. Its function is to produce seminal fluid and spermatophores (sperm-containing capsules).

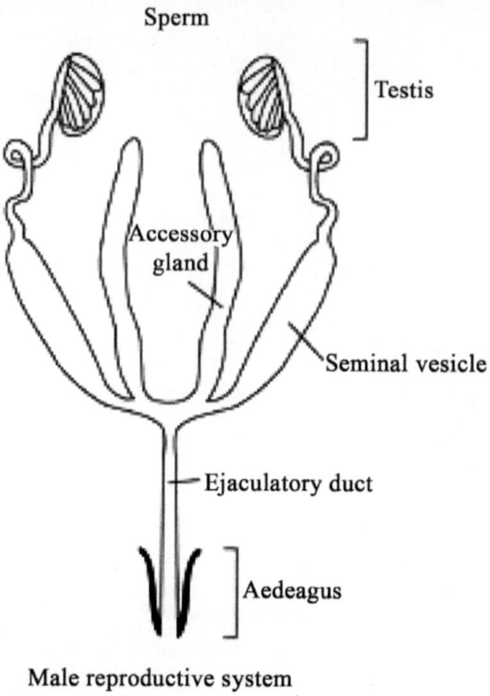

Male reproductive system

10

To Study the Characters of Orders - Orthoptera, Dictyoptera, Odonata, Isoptera, Thysanoptera, Hemiptera, Lepidoptera, Coleoptera, Hymenoptera, Diptera and their Families of Agricultural Importance

Requirements: Agriculturally important insects of each order

Orthoptera

Examples: Grasshopper, cricket, etc.

Metamorphosis: Paurometabolous

Mouthparts: Chewing

Significance: Variable or Pest

Characteristics

- They are medium to large-sized insects.
- Antenna is filiform.
- Mouthparts are mandibulate.
- Prothorax is large. The pronotum is curved, ventrally covering the pleural region.
- Hindlegs are saltatorial
- Forewings are leathery, thickened and known as tegmina.
- They are capable of bending without breaking.
- Hindwings are membranous with large anal area.
- They are folded by longitudinal pleats between veins and kept beneath the tegmina.

- Cerci are short and unsegmented.
- The ovipositor is well developed in females.
- Metamorphosis is gradual. In many orthopterans, the newly hatched first instar nymphs are covered by loose cuticles and are called pronymphs. Wing pads of nymphs undergo reversal during development.
- Specialized stridulatory (sound-producing) and auditory (hearing) organs are present. This order is subdivided into two suborders, viz., **Caelifera** and **Ensifera**.

I. **Families of Caelifera**

1. **Acrididae: (Locusts, Grasshoppers)**

- Antenna is short
- Tarsus is three segmented
- Ovipositor is short and horny
- Tympanum is located one on either side of the first abdominal segment.
- Sound is produced by **femoro-alary** mechanism. A row of **peg** like projections found on theinner side of each hind femur is rubbed against the hard **radial vein** of the closed tegmen.
- Locusts are a serious threat to tropical agriculture. They swarm under favourable conditions andmainly feed on grasses, cereals etc.

II. **Families of Ensifera**

1. Tettigonidae : (Katydids, Long horned grasshoppers)

- Antenna is long, slender as long as or longer than the body.
- Tarsus is four-segmented.
- Ovipositor is sword-like.
- Auditory organs are found in the fore tibiae. In each fore tibia, a pair of tympanum is present. The outer tympanum is larger than the inner.

Sound production is **alary type**. A thick region on the hind margin of the forewing (**scraper**) is rubbed against a row of teeth on the stridulatory vein (**file**) present on the ventral side of another forewing which throws the resonant area on the wing (**mirrors**) into vibrations to produce sound.

2. Gryllidae (Cricket)

- Antenna is long.
- Tarsus is four-segmented.
- Ovipositor is slender and needle like.
- Forewings are abruptly bent down to cover the sides of the body
- Hindwings are acuminate. They are produced into a pair of long processes which project beyondthe abdomen.
- Cerci are long and unsegmented
- Auditory organs and stridulatory organs are similar to long horned grasshopper. Males stridulateduring night. They produce a shrill chirping noise.
- *Gryllus* sp. It is house hold pest.

3. Gryllotalpidae (Mole crickets)

- They are brown coloured insects found inside the burrows. Eyes are reduced.
- Pronotum is elongate, ovate and rounded posteriorly.
- Forelegs are fossorial. Tibiae are expanded and digitate.
- Hindwings are extended beyond the tegima as a pair of processes
- Special stridilatory structures are absent. A humming sound is produced by rubbing the forewings.
- A pair of tympanum is found on the order surface of the tibiae.
- Ovipositor is vestigeal.
- Mole crickets burrow into the soil and feed on tender roots of growing plants. *Gryllotalpa Africana* is a pest on stored potatoes.

Dictyoptera

Examples: Cockroach, praying mantis, etc.

Metamorphosis: Holometabolous

Mouthparts: Chewing

Characteristics

- Head is hypognathous.
- Antenna is filiform.
- Mouthparts are chewing type.
- Tarsus is five segmented.
- Forewings are more on less thickened, leathery with a marginal costal vein. They are calledtegmina.
- Hindwings are large, membranous and folded fan like and kept beneath the forewings.
- Cerci are short and many segmented.
- Eggs are contained in anootheca.

There are two important families viz., **Blattidae** and **Mantidae**

S.No.	Blattidae	Mantidae
1.	Head is not mobile in all directions	Head is mobile in all directions
2.	Head is hidden by the pronotum	Pronotum does not cover the head
3.	Two fenestrae (degenerated ocelli sensitive tolight) occur in the place of ocelli	Three ocelli are present
4.	Pronotum is shield like	Pronotum is elongate
5.	Legs are cursorial and are adapted for running	Forelegs are raptorial middle and hind legs suited for walking
6.	Gizzard is powerfully armed with chitinousteeth to grind food	Chitinous teeth are absent in gizzard
7.	Female does not devour the male during mating	Often (but not always) devours the male during mating
8.	Eggs are laid inside a chitinous ootheca	Eggs are enclosed in filled solidfied foamOotheca is not chitinous.
9.	Nymphs are not cannibalistic	Nymphs are cannibalistic
10	No mimicry is found	Mimic leaves and flowers
11.	Omnivorous	Carnivorous
12.	Found in household, dead wood, litter etc.	Found mostly outdoors
13.	Economic importance: They feed on food stuff, clothes and paper. They impart a foulsmell to the food by contaminating with excreta. Hence they are harmful.	Economic importance: They are predators onmoths, flies, grasshoppers, cater- pillars, etc. Hence they are beneficial.
14.	Important species: American cockroach *Periplanata americana*	Important species: *Mantis religiosa*

Odonata

Examples: Dragonflies, damselflies etc.

Metamorphosis: Hemimetabolous

Mouthparts: Chewing

Significance: Beneficial

Characteristics

- Medium to large-sized insects
- They are attractively colored
- Head is globular and constricted behind into apetiolate neck.
- Compound eyes are large.
- Three ocelli are present
- Mouthparts are adapted for biting. Mandibles are strongly toothed Lacinia and galea are fused to form mala which is also toothed.
- Wings are either equal or sub-equal, membraneous; venation is net work like with many cross veins. Wings have a dark pterostigma towards the costal apex. Sub costa ends in nodus. Wing flexing mechanism is absent.
- Legs are antero ventrally placed. They are suited for grasping, holding and conveying the prey to the mouth. Spinose femora and tibiae are useful for holding the prey. Forward shift of leg attachments allow easy transfer of prey items to mouth in flight. Legs are held in such a way that a basket is formed into which the food is scooped.
- Abdomen is long and slender. In males, gonopore is present on the ninth abdominal segment.
- But the functional copulatory organ is present on the second abdominal sternite. Before mating sperm are transferred to the functional penis. Cercus is one segment.
- Metamorphosis is incomplete with three life stages. The naiad is aquatic. Labium is greatly elongated, jointed and bears two hooks at apex. It is called a mask. It is useful to capture the prey.
- There are two sub-orders. Dragonflies are classified under **Anisoptera** and damselflies are grouped under **Zygoptera.**

S.N.	Anisoptera (Dragonflies)	Zygoptera (Damselflies)
1.	Strong fliers	Weak Fliers
2.	Wings are unequal, Hindwings arebasally broader than forewings	Equal
3.	Wings are broadly attached to theabdomen	Wings are petiolated and narrowly attached
4.	Venation is not similar in both forewings and hind wings.	Venation is identical in both the wings.
5.	Wings are spread laterally at rest	Wings are held at an angle above the abdomen
6.	Compound eyes are large and meetmid dorsally (holoptic)	Compound eyes are button like, wide apart(dichoptic)
7.	Male has three abdominal appendages. Two superior and appendages (cerci) and one inferior anal appendage (epi-proct) are present.	Four terminal abdominal appendages are present.A pair of superior anal appendages (cerci) and a pair of inferior anal appendages. (paraprocts) are present.
8.	Oviposition is exophytic	Oviposition is endophytic

Naids		
1.	Stout and robust	Slender and fragile
2.	Gills are internal and found associated with rectum	Three caudal gills are present which are visible externally.
3.	Able to propel themselves by forcibly ejectingwater through anus from rectum	Lack jet propulsion mechanism

Families of Anisoptera (Dragonflies)

Aeshnidae (Darners): These insects are notable for their large size and brilliant blue or green coloration. Includes the common green darner (*Anax junius*).

Libellulidae (Common Skimmers): This is the largest family in the order. It contains manyspecies with dark spots on the wings.

Gomphidae (Clubtails): These dragonflies have the terminal abdominal segments swollen, hencethe common name.

Families of Zygoptera (Damselflies)

Calopterygidae (Broad winged Damselflies): The wings of these insects are shaped like the seedsof a maple tree.

Coenagrionidae (Narrow winged Damselflies): Small, delicate insects. The body is usually blackwith blue markings. At rest, the wings are held together over the back.

Lestidae (Spread winged Damselflies): These damselflies rest with the body nearly vertical andthe wings partly outspread.

Isoptera

Examples: Termite

Metamorphosis: Hemimetabolous

Mouthparts: Chewing

Significance: Pest

Characteristics

- Pale, elongated bodies, and are sometimes called –white ants.
- Reproductive individuals have two pairs of membranous wings, all of equal length.
- Termites shed their wings after mating.
- Have chewing mouthparts.
- Antennae are roughly the length of their heads

Major Families

1. **Rhinotermitidae (Subterranean termites):** These insects build nests in the soil and generally infestwood that is in contact with the ground.
2. **Hodotermitidae (Rotten wood termites):** Generally found inhabiting moist wood. Contact with thesoil is not a requirement.
3. **Kalotermitidae** (Dry wood and damp wood termites): These insects nest in the wood itself and do notrequire contact with the soil.
4. **Termitidae:** This is the largest family of termites worldwide.

Thysanoptera

Examples: Thrips

Metamorphosis: Hemimetabola

Mouthparts: Rasping - sucking

Significance: Variable

Characteristics

- They are minute, slender, soft bodied insects.
- Mouthparts are rasping and sucking. Mouth cone is formed by the labrum and labium together with basal segments of maxillae. There are three stylets derived from two maxillae and left mandibles. Right mandible is absent. Hence mouth parts are asymmetrical.
- Wings are either absent or long, narrow and fringed with hairs which increase the surface area. They are weak fliers and passive flight in wind is common.
- Tarsus is with one or two segments. At the apex of each tarsus a protrusible vesicle is present.
- Abdomen is often pointed. An appendicular ovipositor may be present or absent.
- Nymphal stage is followed by prepupal and pupal stages which are analogous to the pupae of endopterygote insects.

Major Families of Thysanoptera

1. **Aeolothripidae** (Predatory Thrips): mostly beneficial species.

   Thripidae (Common Thrips): herbivores, including many pests such as the flower thrips (*Frankliniella tritici*), and the soybean thrips (*Sericothrips variabilis*).

2. **Phlaeothripidae (Tube-tailed thrips):** largest family in the order; contains numerous pests as well as a few beneficial species (e.g., the black hunter, *Leptothrips mali*).

Hemiptera

Examples: True bugs, whiteflies, plant hoppers etc.

Metamorphosis: Hemimetabolous

Mouthparts: Piercing-sucking

Significance: Pest

Characteristics

- Head is opisthognathous.
- Mouthparts are piercing and sucking type.

- Two pairs of bristle-like stylets which are the modified mandibles and maxillae are present. Stylets rest in the grooved labium or rostrum. Both labial palps and maxillary palps are atrophied.
- Mesothorax is represented dorsally by the scutellum.
- Forewings are either uniformly thickened throughout or basally coriaceous and distally membranous.
- Cerci are always absent.
- Metamorphosis is usually gradual; but rarely complete.
- Alimentary canal is suitably modified to handle liquid food.(filter chamber)
- Salivary glands are universally present,
- Extra-oral digestion is widespread.
- Abdominal ganglia fused with thoracic ganglia.

There are two suborders viz., Heteroptera and Homoptera.

S.No.	**Heteroptera (Hetero-different; ptera-wing)**	**Homoptera (Homo-uniform; ptera-wing)**
1.	Head is porrect or horizontal	Head is deflexed
2.	Bases of the forelegs do not touch the head	Bases of the forelegs touch the head
3.	Beak arises from the anterior part of the head	Beak arises from the posterior part of the head
4.	Gular region of the head (mid ventral sclerotized part between labium and foramen magnum) well defined.	Gular region not clearly defined
5.	Pronotum usually greatly enlarged.	Pronotum is almost alway small and collar-like.
6.	Scutellum (triangular plate found between the wing bases) well developed	Scutellum not well developed.
7.	Forewings heavily sclerotized at the base andthe apical half is membranous (Hemelytra)	Forewings are of uniform texture. They arefrequently harder than hind pair.
8.	Wings are held flat over the the back at rest and the left and right side overlap on the abdomen.	Wings are held roof-like over the back andwings do not overlap.
9.	Honey dew secretion uncommon	Honey dew secretion common
10.	Repungnatorial or odori-ferous or scent glands present.	Wax glands usually present.
11.	Both terrestrial and aquatic	Terrestrial.
12.	Herbivorous, predaceous or blood sucking.	Herbivorous.

Important families of Heteroptera

1. **Cimicidae (Bedbugs)**
   - Body is dorso ventrally flattened so that it can hide in cracks and crevices. Body is oval in outline.
   - It is dull reddish brown.
   - Thorax is deeply notched in front to receive the short head up to bulging eyes.
   - Hemelytra is short and reduced to scale like pads.
   - Hind wings are completely atrophied.
   - Stinkglandsarelocatedinthedorsalsurfaceoffirstthreeabdominalsegments.
   - Male bed bugs pierce the integument of the female and inject the sperm into the hemocoel during copulation (Haemocoelic or traumatic insemination).
   - Beg bugs hide in crevices of beds, furniture, etc., during the day and emerge at night to seek a blood meal. They are blood-sucking ectoparasites on birds and mammals.
   - They are known for their irritating bite. *Cimex lectularis and Cimex hemipterous* are two important species affecting man in temperate and tropical conditions respectively

2. **Lygaeidae (Seed bugs or Chinch bugs)**
   - Cuneus is absent in hemelytra.
   - Membrane has fewer regular veins (4-5veins) arising from atransverse basal vein.e.g. Dusky cotton bug *Oxycarenus hyalinipennis* nymphs and adults suck the sap from seeds of injured or already opened bolls and reduce the seed quality.

3. **Pyrrhocoridae (Red bugs or Stainers)**
   - They are elongated oval bugs.
   - They show warning coloration. They are brightly marked with red and black.
   - Membrane has more branched veins and cells. e.g. Cotton stainer *Dysdercu s cingulatus*.

- Feeding injury caused by these bugs leads to contamination by the fungus *Nematospora* resulting in yellowish brown discoloration of the lint. Coreidae (Squash bugs or leaf-footed bugs).
- Membrane with many branching veins arising from a transverse basal vein.
- Stink glands are found inside the meta thorax and gland openings are found on the sides of the thorax between the middle and hind coxae. They emit a bad odor. Hind tibia and tarsia are expanded and leaf-like.
- The edge of the abdomen is raised and the wings lie in a distinct depression. e.g. Pod bug, *Riptortus pedestris* nymphs and adults suck the sap from pods of pulses.

4. **Pentatomidae (Stink bugs or Shield bugs)**
   - Antenna is five segmented.
   - Scutellum is prominent and shield like.
   - Adults and nymphs produce a disagreeable odour from stink glands located in metathorax and abdomen respectively.
   - Some are phytophagous and some are predaceous. e.g. Green stink bug *Nezara viridulais* a pest on millets.

Important families of Homoptera

1. **Cicadidae (Cicadas)**
   - Males have sound producing organs at the base of the abdomen. Sound producing organs consist of a pair of large plates, the opercula covering the cavity containing structures producing sound. In the anterior part of the cavity benath each operculum is a yellowish membrane. A shinning mirror is located in the posterior part of the cavity. In the lateral wall of the cavity is an oval shaped ribbed structure, the tymbal. These are vibrated by strong muscles to produce sound. Each species has a characteristic song. Tympanum is present in both in sexes.
   - Wings are transparent.
   - Eggs are inserted into the tree twigs by the female.
   - Nymphs drop to the ground, enter the soil and feed on root sap.
   - Anterior femora of the nymph is thickened with spines beneath and are suited for digging thesoil.
   - Life cycle of periodical cicada lasts for 13-17 years.

2. **Cicadellidae (Leaf hoppers or Jassids)**
   - Elongate insects with a wedge shaped body.
   - Attractively coloured.
   - Hind tibiae have a double row of spines.
   - Ovipositor is well suited for lacerating the plant tissue.
   - Nymphs and adults have the habit of running sidewise.
   - They suck the plant sap and transmit diseases. eg. Green leaf hopper *Nephotettix virescens* transmits tungro disease in rice.

3. **Delphacidae (Plant hoppers)**
   - Large mobile flattened spur is present at the apex of hind tibia e.g. Brown plant hopper *Nilaparvata lugens* causes hopper burn, transmits viral disease in rice Lophopidae
   - Head is produced into a snout.
   - Hind trochanter is directed backward.
   - Hind basitarsus is moderately long. e.g. Sugarcane leaf hopper *Pyrilla perpusilla* nymphs and adults suck the sap and reduce the quality and quantity of cane juice. Aleurodidae (Whiteflies).
   - Minute insects that superficially resemble tiny moths.
   - Wings are opaque and dusted with mealy white powdery wax. Wing venation is much reduced
   - Vasiform orifice is present in the last abdominal tergite. It is a conspicuous opening provided with an operculum. Beneath the operculum, there is a tongue-like organ termed lingula. The anus opens at the base of the lingula through which honeydew is excreted in large amounts.
   - Immature instars are sessile, scale-like, with a waxy covering.
   - Metamorphosis approaches the holometabolous type due to the presence of a quiescent stage prior to the emergence of adults. e.g. Cotton whitefly *Bemisia tabaci* transmits vein-clearing disease in bhendi.

4. **Aphididae (Aphids or Plant lice or Greenflies)**
   - Body is pear shaped
   - Both apterous and alate forms are found.

- A pair of cornicles or siphonculi or wax tubes is present in the dorsum of fifth or sixth abdominal segments which secretes wax like substance.
- They excrete copious amount of honey dew on which ants feed and sooty mould fungus grows.
- Aphids are known for their extraordinary fecundity, short life cycle and parthenogeniticreproduction. Life cycle is highly complex and it involve alternation of generation.
- They feed on plant sap and disseminate plant diseases. e.g. Cotton aphid *Aphis gossypii*.

5. **Coccidae (Scale insects or Soft scales)**

- Sexual dimorphism is present.
- **Male:** They are gnat like, with long antennae, lateral eye and vestigeal mouth parts.
- Mesothorax is enlarged bearing one pair of wings with on or two veins. Hind wings are reduced to halters. A quiescent stage is present in the life history.
- **Female:** Body segmentation is indistinct. Body wall naked and covered with a waxy coating.They are wingless, legless and suck the plant sap.
- The first instar nymph is active and is known as crawler which moults and becomes legless. e.g.Coffee green scale *Coccus viridis*.

6. **Pseudococcidae(Mealy bugs)**

- Body is elongate oval in shape.
- Body segmentation is distinct.
- Body is covered by long radiating thread of mealy secretion.
- Functional legs are present in all instars.
- Wings are absent. e.g. Coconut mealy bug, *Pseudococcus longispinus*.
- Nymphs and adults suck the sap and affect the growth of spindle leaf.

Lepidoptera

Examples: Butterfly and moth

Metamorphosis: Holometabolous

Mouthparts: Siphoning, chewing

Significance: Pest, beneficial, or variable

Characters

Body: Body, wings and appendages are densely clothed with overlapping scales, which give colour, rigidity and strength. They insulate the body and smoothen air flow over the body.

Mouthparts: Mouthparts in adults are of **siphoning type**. Mandibles are absent. The **galeae of maxillae are greatly elongated** and are held together by interlocking hooks and spines. The suctorial proboscis is coiled up like a watch spring and kept beneath the head when not in use.

Wings: Wings are membranous and are covered with overlapping **pigmented scales**. Forewings are larger than hind wings. Wings are coupled by either frenate or amplexiform type of wing coupling.

Larvae: Larvae are **polypod-eruciform** type. Mouthparts are adapted for chewing with strong mandibles. There are three pairs of five segmented thoraciclegs ending in claws. Two to five pairs of fleshy unsegmented **prolegs** are found in the abdomen. At the bottom of the proleg, **crochets** are present.

Pupae: Pupa is generally **obtect**. It is either naked or enclosed in a cocoon made out of soil, frass, silk or larval hairs.

Sub-orders: Ditrysia and Monotrysia

Most of the lepidoptern insects **(97 %)** are grouped under the suborder **Ditrysia** in which the female insects have two pores **i.e., the copulatory pore** is located in the **8th** abdominal sternite and the egg pore in the **9th** abdominal sternite. Remaining insects are grouped under the **suborder Monotrysia** in which the female insects have **one pore**.

Butterfly Family

1. **Papilionidae (Swallotail Butterfly):** They are often large and brightly coloured (Figure: on cover page). Prothoracic legs have tibial epiphysis. In many species hind wings has tail like prolongation. Amplexiform type of wing coupling is present. Larval body is either smooth or with tubercles. Retractile Osmeteria are present on the prothoracic tergum of the caterpillar. e.g. Citrus butterfly, *Papilio demoleus*

2. **Pieridae (Whites and sulphurs):** They are white or yellow or orange coloured with black markings. Larva is green, elongate and covered with fine hairs. Larval body segments have annulets. e.g. Cabbage white butterfly,*Pieris brassicae.*

Moth Family

1. **Arctidae (Tiger moth):** Wings are conspicuously spotted or banded. They are nocturnal and attracted to light. Larva is either sparsely hairy or densely hairy (wooly bear). e.g. Spotted boll worm, *Earias vitella*, Sunhemp caterpillar, *Utetheisa pulchella*
2. **Bombycidae (Silk worm moths):** Antenna is bipectinate. Larva is either with tuft of hairs or glabrous with medio dorsal horn on 8th abdominal segment. Pupation occurs in dense silken cocoon. e.g. Mulberry silkworm, *Bombyx mori*
3. **Gelichidae (Paddy moths):** Forewings trapezoidal and narrower than hind wing. Caterpillars bore into the seeds tubers and leaves. e.g. Cotton pink boll worm, *Pectinophora gossypiella, Angumous grainmoth, Sitotroga cerealella*, Potato tuber moth, *Pthoremea operculella*
4. **Noctuidae (Noctua moths):** They are medium sized, stoutly built moths. They are nocturnal and attracted to lights. Labial palpi is well developed. All crochets on the larval Prolegs are of same size and arranged in semi circle. Some larvae are semiloopers. They have either 3 or 4 pairs of prolegs. Larvae attack the plants during night. Larvae of some species remain concealed beneath the surface of the ground or litter during day and feed on plants during night. They often cut small seedlings close to the ground and hence they are called cut worms. e.g. Tobacco cut worm, *Spodoptera litura*
5. **Pyraustidae/ Pyralidae:** Proboscis is vestigial in many species. Labial palp is snout like. Larval habit varies. It may live among aquatic plants and bore into the stem or remain in silken web among spun up plants parts. Some larvae are aquatic and gill breathing. e.g. Rice stem borer, *Scirpophaga incertullus.*

Coleoptera

Examples: Beetles

Metamorphosis: Holometabolous

Mouthparts: Chewing

Significance: Pest, beneficial, and variable

Body: Minute to large sized insects

Characteristics

- They are minute to large sized insects.
- Antenna is usually 11 segmented.
- Mouthparts are chewing type.
- Mandibles are short with blunt teeth at the mesal face in phytophagous group.
- In predators the mandibles are long, sharply pointed with blade like inner ridge.
- In pollen feeders teeth are absent and the mandibles are covered with stiff hairs.
- Prothorax is large, distinct and mobile.
- Mesothorax and metathorax are fused with the first abdominal segment.
- Forewings are heavily sclerotised, vein less and hardened. They are called elytra.
- Forewings do not overlap and meet mid-dorsally to form a mid-dorsal line. It is not used for flight. They serve as a pair of convex shields to cover the hind wings and delicate tergites of abdomen.
- Hindwings are membranous with few veins and are useful in flight. At rest they are folded transversely and kept beneath the elytra. In some weevils and ground beetles the forewings arefused and hind wings are atrophied.
- A small part of the mesothorax known as scutellum remains exposed as a little triangle between the bases of elytra.
- Cerci and a distinct ovipositor are absent.
- Metamorphosis is complete. Larvae are often called grubs.
- Pupae are usually exarate and rarely found in cocoons.

Sub-orders: Adephaga (predators/devourers) and Polyphaga (eaters of many things). **Families of predators:** Cicindelidae (Tiger beetle), Carabidae (Ground beetle), Dytiscidae (True water beetle), Gyrinidae (whirligig beetle), Coccinellidae (Ladybird beetle), **Lampyridae (Firefly, glow worm)**

Families of scavengers: Scarabaeidae (Scarabs, dung beetle), Hydrophilidae (water scavenger beetle) Families of stored product pests:

Anobiidae (Wood worm/ wood borer) e.g. Cigarette beetle- *Lasioderma serricorne* Bostrychidae (Grain borer) e.g. Lesser grain borer - *Rhizopertha dominica*

Families of crop pests:

1. Apionidae e.g. Sweet potato weevil, *Cylas formicarius*, a pest both in the field and instorage. Head is produced into snout. Antennae are not elbowed. Grubs are apodous.

2. Cerambycidae (Longicorn beetles/ Longhorn beetles)

e.g. Mango stem borer, *Batocera rufomaculata*

3. **Curculionidae (Weevils/ snout beetles)**

Minute to large sized insects. Frons and vertex of the head produced into snout, which is cylindrical and in some species larger than the beetle itself. Mouthparts (mandible and maxilla) are present at the tip of the snout, It is useful to feed on internal tissues of the plant and provide a place for egg laying. Antenna is geniculate and usually found in the middle of the snout. Grubsare apodous and acephalous. Weevils are important crop pests during both in field and in storage. e.g. Coconut red palm weevil, *Rhynchophorus ferrugineus*

3. **Galerucidae / Chrysomelidae (Pumpkin beetle)**

Antenna are closely approximated. Third tarsomere is deeply bilobed. Larvae are root feeders. Adults bite holes on plants. Red pumpkin beetles, *Raphidopalpa foveicollis* **Melolonthidae (Chaffer beetle, June beetle, White grub)**

They are stout beetles with glossy surface. Head is small. Labrum is well sclerotized. Adults are attracted to lights. They feed on tree foliage during night and hide in soil during day time. Larvae are scarabaeiform and root feeders. e.g. Groundnut white grub, *Holotrichia consanguinea*, a serious pest under rainfed conditions.

Families of Predators

Coccinellidae (Lady bird beetle)

They are hemispherical insects. The body is convex above and flat below. The body appearance resembles like a split pea, head is small, turned down ward and received into a prominent notch of prothorax. Elytra is strongly convex, brightly coloured and variously spotted. Grubs are Campodeiform and spiny. The last larval skin either covers the pupa or gets attached to its anal end.

Except the genus Epilachna others are predators on aphids, scales, mites and whiteflies. e.g. *Coccinella septempunctata, Coccinella vigintioctopunctata.*

Hymenoptera

Examples: Ant, bee, wasp

Metamorphosis: Holometabolous

Mouthparts: Chewing,

Chewing-lapping

Key Membranous wings, front pair larger than rear

Significance: Pest, beneficial, or variable

Characteristics

- Mouthparts are primarily adapted for chewing. Mandibles are very well developed. In bees bothlabium and maxillae are integrated to form the lapping tongue.
- Thorax is modified for efficient flight. Pronotum is collar like. Meso thorax is enlarged. Metathorax is small. Both prothorax and metathorax are fused with mesothorax.
- Wings are stiff and membranous. Forewings are larger than hind wings. Wing venation is reduced. Both forewings and hindwings are coupled by a row of hooklets (hamuli) present on the leading edge of the hind wing.
- Abdomen is basally constricted. The first abdominal segment is called propodeum. It is fused with metathorax. The first pair of abdominal spiracles is located in the propodeum. The second segmentis known as pedicel which connects the thorax and abdomen. Abdomen beyond the pedicel is calledgaster or metasoma.
- Ovipositor is always present in females. It is variously modified for oviposition or stinging or sawing or piercing plant tissue.
- Metamorphosis is complete. Often the grub is apodous and eucephalous. Larva is rarely eruciform. Pupa is exarate and frequently enclosed in a silken cocoon secreted from labial glands.
- Sex is determined by the fertilization of the eggs. Fertilized eggs develop into females and males are produced from unfertilized eggs. Males are haploid and females diploid.

This order is subdivided into two suborders: **Symphyta and Apocrita**

Suborder Symphyta- Abdomen is broadly joined with thorax, stemmata present, ovipositor issaw like and suited for piercing plant tissues, habits are phytophagous ,e.g. sawflies and horntails

Family: Tenthrinidae (Saw flies)

They are wasp like insects. Abdomen is broadly joined to the thorax. Ovipositor is sawtoothed and suited for slicing the plant tissue. Larva is eruciform. It resembles a lepidopteran caterpillar. It has 1 paired celli, papillae (reduced antenna), 3 pairs of thoracic legs and 6-8pairs of abdominallegs. Prolegs lack chrochets. They are external feeder on the foliage. Larvae while feeding usually have posterior part of the body coiled over the edge of the leaf. Mustard Sawfly, *Athlia lugens proxima* is a defoliator of mustard and cruciferous vegetables.

Suborder Apocrita- Abdomen is petiolated, Stemmata are absent, Legs are absent, ovipositor is not saw like and Suited for piercing or stinging, they are generally parasitic.

Family: Apidae (Honey bee)

Body is covered with branching or plumose hairs. Mouth parts are chewing and lapping type. Mandibles are suited for crushing and shaping wax for building combs. Legs are specialized for pollen collection. Scopa (pollen basket) is present on hind tibia. They are social insects with 3 castes viz. Queen, drone and workers. Division of labour is noticed among honey bees. Indian honey bee, *Apis indica*

Formicidae (Ants)

They are common widespread insects. Antennae are geniculate. Mandibles are well developed. Wings are present only in sexually mature forms. Petiole may have 1 or 2 spines. They are social insects with 3 castes viz. Queen, males and workers. Workers are the sterile females and theyform the bulk of the colony. Exchange of food materials between adults and immature insects is common. After a mating flight queen alone finds a suitable nesting site. Many species have associated symbiotic relationship with Homopteran insects.

Braconidae (Braconid wasp)

They are small, stout bodied insects. Fore wing has one recurrent vein. Petiole is neither curved nor expanded at the apex. Gaster is sessile or sub sessile. Abdomen is as long as head and thorax together. They parasitize lepidopteran larvae commonly. They are gregarious parasites. In many species, Poly embryony is observed. *Bracon brevicornis* is mass multiplied and released for the control of black headed coconut caterpillar.

Diptera

Examples: Fly midge, mosquito, etc.

Metamorphosis: Holometabolous

Mouthparts: Piercing-sucking, cutting-sponging, sponging, chewing, cutting-lappingSignificance: Pest, beneficial, or variable

Characteristics

- They are small to medium sized, soft bodied insects.
- The body regions are distinct.
- Head is often hemispherical and attached to the thorax by a slender neck.
- Mouthparts are of sucking type, but may be modified.
- All thoracic segments are fused together. The thoracic mass is largely made up of mesothorax. Asmall lobe of the mesonotum (scutellum) overhangs the base of the abdomen.
- They have a single pair of wings.
- Forewings are larger, membranous and used for flight.
- Hindwings are highly reduced, knobbed at the end and are called halters. They are rapidly vibrated during flight. They function as organs of equilibrium. Flies are the swiftest among all insects.
- Metamorphosis is complete. Larvae of more common forms are known as maggots. They are apodous and acephalous. Mouthparts are represented as mouth hooks which are attached to internalsclerites. Pupa is generally with free appendages, often enclosed in the hardened last larval skin called puparium. Pupa belongs to the coarctate type.

Families of agricultural Importance:

Syrphidae (Horse flies, Flower flies)

They are brightly coloured and brilliantly stripped. A vein like thickening is present in between the radius and median in the forewing. Abdomen has distinct black and yellow markings. Maggots prey on soft bodied insects especially aphids. Adults are excellent fliers. They hoverover flowers. They feed on pollen and nectar. They aid in pollination.

Tephritidae (Fruit flies)

Sub costa bends apically and fades out. Wings are spotted or banded. Female has a sharp and projecting ovipositor. Maggots can hop. They are highly destructive to fruits and vegetables. Cucurbit fruit fly, *Dacus cucurbitae*. Tachinidae (Tachinid flies).

11

To Study the Insecticides and Their Formulations

Requirements: Insecticides of different types

Insecticide chemicals in their "raw" or unformulated state are not usually suitable for pest control. These concentrated chemicals and active ingredients may not mix well with water, may be chemicallyunstable, and may be difficult to handle and transport. For these reasons, manufacturers add inert substances, such as clays and solvents, to improve application effectiveness, safety, handling, and storage. Inert ingredients do not possess Insecticidal activity and are added to serve as a carrier for the active ingredient. Manufacturers will list the percentage of inert ingredients in the formulation or designate them as "other ingredients" on their labels. There are several inert substances, such as petroleum distillates and xylene, which will have a specific statement identifying their presence in the formulation. The mixture of active and inert ingredients is called a pesticide formulation. Thisformulation may consist of:

- The pesticide active ingredient that controls the target pest.
- The carrier, such as an organic solvent or mineral clay.
- Adjuvants, such as stickers and spreaders
- Other ingredients, such as stabilizers, safeners, dyes, and chemicals that improve or enhance pesticidal activity

Types of Formulations

- Emulsifiable concentrates
- Wettable powders
- Dusts
- Aerosols
- Granular formulations
- Baits
- Micro encapsulated products

Emulsifiable Concentrates (EC or E)

An emulsifiable concentrate formulation usually contains a liquid active ingredient, one or more petroleum-based solvents (which give EC formulations their strong odor), and an agent—known as an emulsifier—that allows the formulation to be mixed with water to form an emulsion. Upon mixing with water, they take on a "milky "appearance.

Most ECs contain between 2.5% and 75% active ingredient. ECs are among the most versatile formulations. They are used against agricultural, ornamental and turf, forestry, structural, food processing, livestock, and public health pests. They are adaptable to many types of application equipment including portable sprayers, hydraulic sprayers, low-volume ground sprayers, mist blowers, and low-volume aircraft sprayers.

Wettable Powders

Wettable powders are dry, finely ground formulations that look like dusts. They usually must be mixed with water for application as a spray. Wettable powders contain 5%–95% active ingredient by weight. The particles do not dissolve in water. They settle out quickly

unless constantly agitated to keep them suspended. Wettable powders are one of the most widely usedpesticide formulations. They can be used for most pest problems and in most types of spray equipment where agitation is possible.

Dust

Most dust formulations are ready to use and contain a low percentage of active ingredients (usually 10% or less by weight), plus a very fine, dry inert carrier made from talc, chalk, clay, nut hulls, or volcanic ash. The size of individual dust particles varies.

A few dust formulations are concentrates and contain a high percentage of active ingredients. These concentrates are mixed with dry inert carriers before applying.

Dusts are always used dry and can easily drift to non target sites. They are widely used as seed treatments and sometimes for agricultural applications. In structures, dust formulations are used in cracks and crevices and for spot treatments to control insects such as cockroaches. Insects ingest poisonous dusts during grooming or absorb the dusts through their outer body covering. Dusts also are used to control lice, fleas, and other parasites on pets and livestock. Ready-to-use aerosols are usually small, self-contained units that release the pesticide when the nozzle valve is triggered. The pesticide is driven through a fine opening by an inert gas under pressure, creating fine droplets. These products are used in greenhouses, in small areas inside buildings, or in localized outdoor areas. Commercial models, which hold 5–10 pounds of pesticide, are usually refillable.

Aerosols

These formulations contain one or more active ingredients and a solvent. Most aerosols contain a low percentage of active ingredients. There are two types of aerosol formulations: the ready-to-use type commonly available in pressurized, sealed containers and those products used in electric- or gasoline-powered aerosol generators that release the formulation as a "smoke" or "fog."

Granular formulations

Granular formulations are similar to dust formulations except granular particles are larger and heavier. The coarse particles are made from materials such as clay, corncobs, or walnut shells. The active ingredient either coats the outside of the granules or is absorbed into them. The amount of active ingredient is relatively low, usually ranging from 1 to 15 percent by weight.

Granular pesticides are most often used to apply chemicals to the soil to control weeds, fire ants, nematodes, and insects living in the soil or for absorption into plants through the roots. Once applied, granules release the active ingredient slowly. Some granules require soil moisture to release the active ingredient. Granules are used in agricultural, structural, ornamental, turf, aquatic, and public health (biting insect) pest control operations.

Baits

A bait formulation is an active ingredient mixed with food or another attractive substance. The bait either attracts the pests or is placed where the pests will find it. Federal regulations require that certain rodenticide baits must be contained in tamper-resistant bait stations.

Pests are killed by eating the bait that contains the pesticide. The amount of active ingredient in most bait formulations is quite low, usually less than 5%.

Baits are used inside buildings to control ants, roaches, flies, other insects, and rodents. Outdoors they sometimes are used to control snails, slugs, and insects such as ants and termites. Their main useis for control of vertebrate pests such as rodents, other mammals, and birds.

Microencapsulated products

Manufacturers cover liquid or dry pesticide particles in a plastic coating to produce a microencapsulated formulation. Microencapsulated pesticides are mixed with water and sprayed in the same manner as other sprayable formulations.

After spraying, the plastic coating breaks down and slowly releases the active ingredient.

12

To Study the Pesticide Appliances and Their Maintenance

Sprayers

Sprayer is a machine used to apply liquid chemicals on plants to control pest and diseases. It can also be used to apply herbicides to control weeds and to spray micro nutrients to enhance plant growth. The main functions of a sprayer are

- Breaking the chemical solution in to fine droplets of effective size.
- Distributing the droplets uniformly over the plants.
- Applying the chemicals with sufficient pressure for positive reaching the plants
- Regulating the amount of liquid applied on plants to avoid excessive application.

A variety of high volume sprayers are available in the market. Almost all types of high volume sprayers have some kind of pump to supply pressurised spray liquid to the hydraulic nozzle which breaks the liquid into spray droplets and throws the spray away from it. The high volume sprayers are both manually operated or power operated type.

Principle: The function of sparyer is to atomize the spary fluid in to small droplets and eject it with some force.

Parts of sprayers: The important parts are tank, agitator, pressure gauge, valves, filters, pressure chamber, hose, spray lance, cut off device, boom and nozzle.

Nozzle body: It is the main component on which other component of a nozzle fit

Swirl plate: It is the part of a cone nozzle which imparts rotation to the liquid passing through it

Spray gun: It is a lance from which spray is readily adjustable during the operation.

Spray boom: It is a spray lance with spray nozzles fitted to a head, mounted at right angles to the lance.

Filter: It is a component to remove suspended matter larger than a predetermined size from fluid. Over-flow pipe: It is a conduit through which excess fluid from a pump is by-passed by the action of a relief valve or pressure regulator.

Relief valve: It is an automatic device to control the pressure of fluid or gas within range a predetermined value.

Pressure regulator: It is an automatic device to control the pressure of fluid or gas within a range of settings.

Cut-off valve: It is a mechanism between the pump and the nozzle to control the flow of liquid from the sprayer. This is operated by hand.

Nozzle disc: It is component containing the final orifice of a nozzle usually a cone nozzle. **Nozzle boss:** It is a lug on spray boom or spray lance to which a nozzle body or cap is screwed. Nozzle tip: It is component containing the final orifice of a nozzle usually a fan nozzle.

Spray lance: A hand-held pipe through which the liquid reaches the nozzle mounted at the free end.

Types Of Nozzle: The three common types of nozzle

a. **Hollow cone nozzle:** This liquid is fed into a whirl chamber through at an gential entry or through a fixed spiral passage to give a rotating motion. The liquid comes out in the form of a harrow conical sheet which then breaks up into small drops. This is used for insecticide and fungicide spraying.

b. **Solid cone nozzle:** This nozzle covers the entire area at small range. The construction is similar to hollow cone nozzle with the addition of an internal jet which strikes the rotating liquid just within the orifice of discharge. The breaking of drop is mainly due to impact. This is used for herbicide spraying.

c. **Fan nozzle:** It is a nozzle which forms narrow elliptical spray pattern. In this type the liquid is forced to come out as a flat fan shaped sheet which is then broken into droplets. This nozzle is mostly used for low pressure spraying.

Types of Sprayers

A. **Manually Operated Hydraulic Sprayers-** In this type, the hydraulic pump directly acts on the spray fluids and discharge it.

a. **Hand syringe:** It is a single acting pump working on the principle of cycle pump. it consists of cylinderin to which the spray fluid is drawn during the suction stroke and delivered during the pressure stroke and discharge through nozzle. It is useful to operate only a small area.

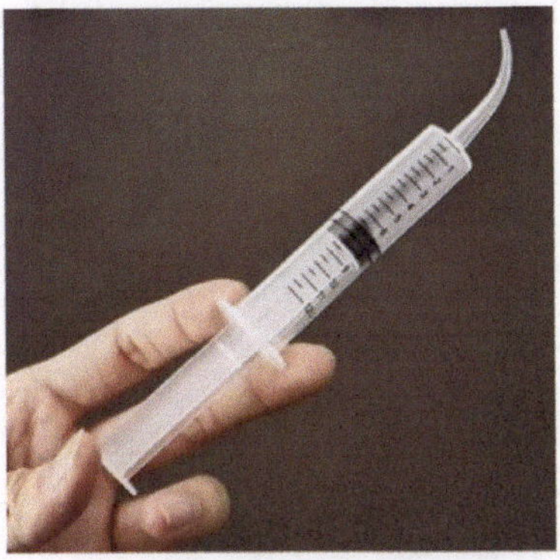

b. **Hand Sprayers:** This is a simple sprayer. It creates hydraulic pressure by forcing spray solution to a nozzle by the direct action of hand pumping. The spray solution is filled in a plastic can (5-10 L) which is usually shoulder slung. A dip-tube draws liquid from the tank due to hand actuation of the plunger. Held by both the hands the piston pump is worked by sliding action. The capacity of this sprayer is about 0.5 acre per day. It is useful for small scale spraying in nursery or kitchen gardens and pot plants.

c. **Bucket or Stirrup Pump Sprayer:** It consist either of a double acting pump with two cylinders or a single acting pump with one cylinder. The other parts of the sprayer are the plunger assembly, foot value assembly, hose, lance and nozzle, a stirrup and an adjustable foot rest. The suction part of the pump is immersed in the spray solution kept on floor in a bucket. The pump is operated by hand by one person while the other person holding the delivery line, trigger cut-off device and lance nozzle sprays pesticide. This sprayer is used both for public health spraying and agricultural spraying purposes. This type of sprayer is useful for spraying small trees. Area covered per day is 0.5 to 0.8 ha.

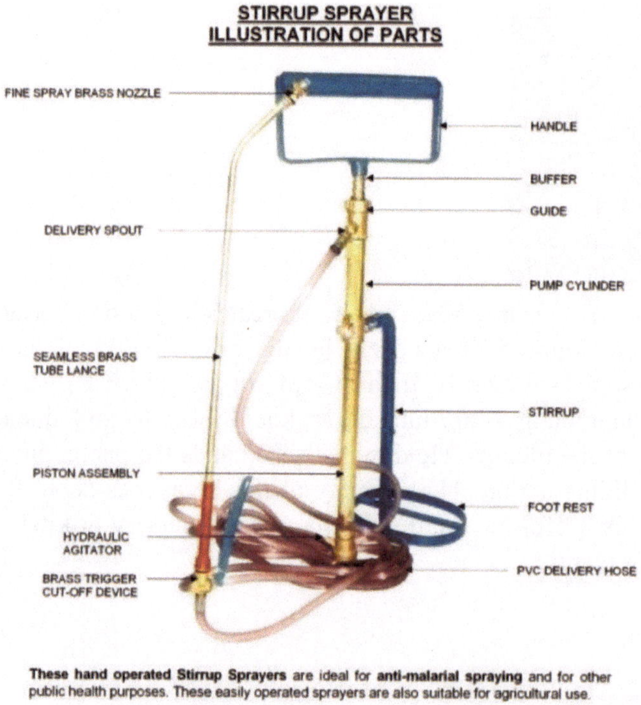

d. **Knapsack Sprayer:** The sprayer is mounded on the back of operator with help of a pair of mounting straps. The pump of the sprayer is actuated by working a hand lever up and down by one hand of the operator and the other hand holds the cut off device for spraying purpose. This sprayer consists of liquid tank, hydraulic pump, operating lever, pressure chamber, agitator, delivery hose, spray lance and nozzle. A bean shaped plastic tank of 14-16 liters capacity is commonly used. It is necessary to operate the hand lever continuously at the rate of 15-20strokes per minute. The normal working pressure is 40 psi. It is user for spraying field crops vegetables and nurseries. The area covered per day is 0.8 to 1 ha.

e. **Rocker Sprayer:** It is very much similar to the foot sprayer. The main difference is the operation of pump. The pump actuation is done by hand of the operator. The sprayer pump mounted on wooden platform is kept on ground and the spray solution is kept in a separate tank or container. It can develop high pressure 10 kg/cm2. For spraying tall trees, an extension bamboo lance can be fitted. The adjustable type hydraulic nozzle (Triple Action Nozzle) is normally used. It can be used for spraying trees and tall field crops. It covers about 1.5 to 2 hectares of area in a day.

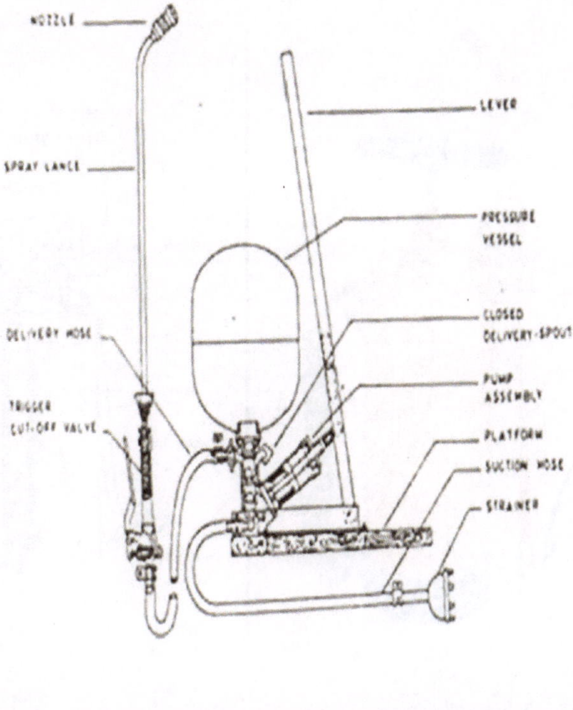

f. **Foot Sprayer or Pedal Pump:** The pump of the sprayer is worked by operating a pedal lever by the foot of the operator. The spray liquid is kept in bucket or container and it is sucked by a suction hose through a filter (strainer) due to piston movement. A suitable ball valve is provided in the piston assembly to serve as suction valve. The liquid from the pump cylinder is then delivered into a pressure chamber where from the pressurized liquid reaches hydraulic nozzle. Minimum two person team is required to work on this machine. Hydraulic pressure of 10 kg/cm2 can be achieved which is necessary to project the jet of spray to tall trees simultaneously from two spray nozzles. The foot operated sprayer is basically for orchard and tree spraying. The design is strong and sturdy. An adjustable type hydraulic nozzle (Tripple Action Nozzle) is generally used which can generate different types of spray patterns viz., fine spray (hollow cone), medium spray and coarse spray (jet).The fine and medium spray are suited for low height orchards, jet spray are necessary for tree spraying. The spray jet can reach height of 15 - 20 feet. For spraying taller trees an extra extension like bamboo lance may be used to gain additional height by 8 - 10 feet. It is difficult to treat field crops by foot sprayers because the sprayer is kept on ground and pesticide solution tank is also kept on ground separately and so movement of the long delivery hose becomes very difficult. About 1 to 1.5 ha area can be sprayed in a day.

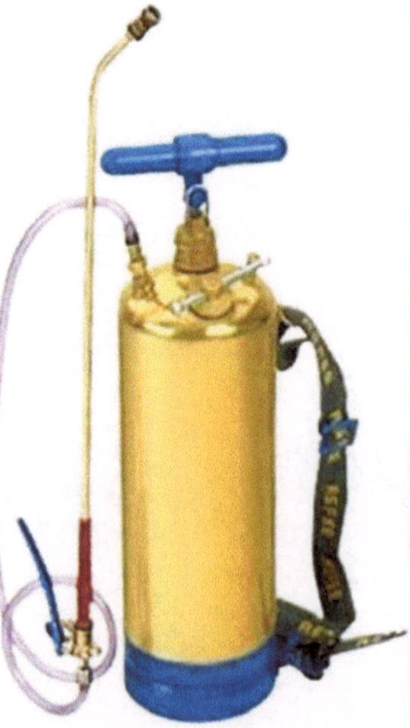

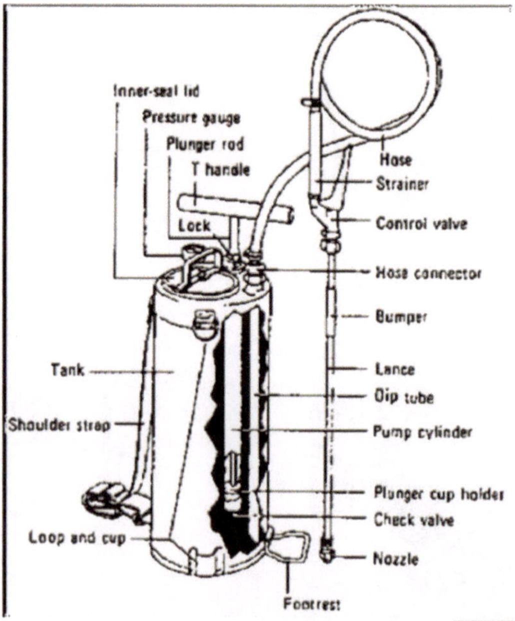

B. Manually Operated air compression Sprayers

These are also known as pneumatic sprayers because air pressure is employed for forcing the liquid though the nozzle for atomization. The containers of these sprayers should not be filled completely with the spray fluid. A part of the container is kept empty so that adequate air pressure can be developed over the spray fluid in the tank. They do not have agitators and hence are not useful spraying materials which settle down quickly.

a. **Pneumatic Hand Sprayer:** The container for the spray fluid also acts as the pressure chamber. An air pump attached to the chamber inside. The inner end of the discharge pipe runs down to the bottom of the container and its outlet terminates in a nozzle is filled about 3/4th of it and the pump is worked force air into the space to build sufficient pressure upon the spray fluid. These sprayers are used extensively in kitchen gardens, in glasshouses and in doors against household insects. The capacity of tank is up to one liter, if used in field it can cover an area of 0.1 ha in a day.

b. **Pneumatic Knapsack Sprayer:** This is similar to compresssion hand sprayer but are used for spraying large quantities of liquids (9-10 Litres). It comprises a tank for holding the spray as well as compressed air, a vertical air pump with a handle, filling hole with a strainer, spray lance with nozzle and release and shut-off devices. The tank is provided a convenient rest with the back of the operator and has shoulder straps that allow it to be carried by him. These sprayers are used against agricultural pests and mosquito control operations. This pump covers an area of about 0.8 to 1.2 ha in a day.

c. **Power Sprayer (Mist blowers cum Duster):** Here the spray fluid is blown out by an air produced in the machine. It consists of chemical tank, fuel tank, carburator, spark plug, engine, blower assembly, delivery system, nozzle system and starter pulley. The power operated spraying system can be converted into a dusting unit by changing certain components. The tank in these is made of a thick polyethylene and has a capacity of 10 liters. The fuel tank capacity is 1.0 to 1.5 liters. It is provided with 1.2 to 3.0 hp petrol engine. This can also be used for dusting provided suitable accessories. The area covered by these sprayers is about 2 ha in a day.

d. **Hand carried, battery operated spinning disc sprayer (Ultra Low Volume Sprayer):** The pesticides are applied as such or with less than 5 litres spray fluid produces fine droplets (80μm).These are light weight sprayers(<3kg) have a rotary atomizer(spinning disc) powered by an enclosed DC motor with a plastic spray head, a liquid reservoir, a handle and a power supply unit.

Liquid is gravity fed from polyethene container screwed in to the spray head moulding and the liquid is flung off by centrifugal force.

e. **Electrodyn Sprayers (EDS):** It is new system of spraying for the controlled droplet application of chemicals (CDA).EDS puts more of active chemical on the target than any other spraying system since the charged particles are attracted to target crop which ensure coverage on the underside of leaves where many pests feed and also there is minimal drift to non target areas. The EDS consists of a spray stick and an unique combination of bottle plus nozzle the bozzle. The spray stick consists of the batteries and a solid state high voltage generator. The bozzle contains ready formulated chemical for immediate application to crops. The pesticide in ULV formulation is used undiluted at a quantity less than 6 liters/ha and usually at 0.5 to2.0 liters/ha for field crops. The droplet size varies from 20-150 micron with ground spraying equipment for ULV spray an area of 5 ha can be covered in a day.

Dusters

The dusting powders are low concentration ready to use type, dry formulations containing 2 to 10% pesticide. The inert material or dry diluents is talc, soapstone, attapulgite, etc., and it is non toxic. The sulphur dust is not diluted with inert material. The dusts are applied at 20 - 50 kg/ha. It should be noted that the application is done in highly concentrated form, as compared to high volume or low volume spraying technique. Therefore, adequate precautions must be taken in handling the dust and during the application in field. The dusters are available both manually operated and power operated models. All dusters consist essentially of a hopper which usually contains an agitator, an adjustable orifice and delivery tubes. A rotary fan or a bellows provides the conveying air.

A. Manually operated dusters

a. **Plunger duster:** They are very simple, low cost machines and useful in a limited way. It consists of a dust chamber, a cylinder with a piston or plunger, a rod and a handle. The field application capacity is low. They hold 200 to 400 g of dust in a chamber into which air is pushed by an adjoining piston type air pump operated by hand. The dust cloud is issued from the discharge outlet. It is useful for small scale use in kitchen garden and in household.

b. **Bellows duster:** This is also a simple design low cost dusting machine. A collapsible bellows pushes air into a dust hopper of 1-2 kg capacity and dust is discharged from the nozzle outlet.

c. **Rotary duster:** This type of duster makes use of a fan or blower to flow large volume of air at high speed. The dust powder is fed into the stream of air and blown from the outlet tube. The fan or blower rotates at high speed by hand cranking handle, which is geared to it. The higher gear-ratio and better blower design provide easy cranking and good volume of air is emitted. The dust hoppers are generally cylindrical and are provided with agitator, feeders and dust metering mechanism. Such rotary dusters are either shoulder slung type or belley mounted type. The shoulder-slung models are better balanced when the dust hoppers are filled. But it becomes inconvenient to operate in crops like sugarcane and cotton. The belley mounted type can be used in such situations. A hand rotary duster can discharge dust powder from 0 – 150 g/min and displace air about one m3/min at 35 RPM. Such machine can treat 1 to 1.5 ha /day.

d. **Power Duster:** These are bigger machines run with the help of engine or electrical motor. Some power dusters are tractor mounted type and are driven by tractor P.T.O. The equipment is mounted on iron frame (stretcher) and can be carried by 2-3 men. The engine/motor drives a centrifugal fan usually via V-belt drive. The engine is petrol/ diesel run and 3 - 5 H. P. The fan displaces 20 m3 air/min or more at 100-250 km/hr air velocity. These dusters are good for large area treatment and suitable for application on tall trees. In this type of duster design, usually the dust powder is not rotated in the fan-case but dust powder is aspirated in the delivery channel by air blast. The dust hopper capacity is 10- 20 kg and dust can bed is charged at a rate of 1 to 8 kg/min. A power duster can cover about 10 ha/day.

e. **Knapsack Duster:** The motorised knapsack sprayer can be converted to a duster by replacing some plastic fittings inside the hopper. Almost all mist blowers have provision of converting them from spraying unit to dusting unit. The two stroke petrol engine runs a blower fan and delivers the air through a hose pipe system. The dust is agitated and lifted by the blast of air in the hopper (2-5kg capacity) and it is fed into the main air hose or a long dusting hose (40-50 ft long polythene perforated hose) can also be attached to knapsack duster. Such an attachment is very good for large area treatment in less time. The dust output can be adjusted from 0 to 1.5 kg/min. The motorised knapsack sprayer- cum-duster unit is therefore useful for both low volume spraying and dusting operation.

Soil Injector

It is also known as soil gun, which consists of a cylindrical tank for the liquid fumigant, a pump barrel and plunger assembly, injector nozzle, thrust handle and injection handle. The hand operated soil injectors have a capacity of 1 to 3 liters and they can cover about 0.5ha in a day. They are used to apply liquid nematicides to kill soil nematodes.

Granule Applicators

They are used to apply granular formulations of pesticides uniformly. These are two types of granular applicators.

I. There is a plastic hopper 1 liter capacity from which the granules flow by gravity to a nozzle.

II. It is a knapsack type with hopper of 10 liters capacity.

Bird Scarer

It produces loud noise at regular interval and used to scare away the birds. It has three essential chambers, a chamber to hold calcium carbide, a smaller chamber placed inside the former to hold water and combustion chamber attached to the main chamber. Water acts with calcium carbide and generates acetylene which explodes producing the noise. The frequency of flow of water into calcium carbide chamber. One kg of calcium carbide is sufficient for working a machine for 24 hours. One bird scarer is sufficient to cover 1 to 2 ha.

Rat Traps

Several types of mechanical devices for trapping rats and mice are used in India. In these traps baits like dry fish are used for attracting these rats. The cage type wooden box with a door closing device and spring board types are the more common ones used in the houses.

13

To Study About the Sampling Techniques for Estimation of Insect Population and Damage

Requirements: Clipboard, Pencils, Hand lens, Pocket knife, insect catching nets, attractants, spread sheet Collection bags and vials, Camera etc.

Sampling techniques are used to draw valid inferences about the population parameters using the sample statistics

Components of an insect sampling program

Knowledge of pest and beneficial insects

Identification

Life cycle and biology

Injury caused

Sample: Sample is a subgroup of the population. It is defined as a collection of individual observations from the population about which inferences are to be made, and is obtained by a specific method.

Sampling: It refers to the process of selecting a portion of the population to represent the entire population.

Sampling gives valuable information

- Detect pest presence
- Detect damaging stage
- Presence of natural enemies
- Populations increasing or decreasing in density (from repeated samples)
- Effectiveness of control measures

Nature of sample

This depends on the insect, its distribution and the stage of the pest to be sampled. The methods are:

Net sweepings	Sight counting
Sudden trapping	Fixed volume or area of earth
Light Trap catch	Crop samples
Water trap	Emergence cages
Suction trap	Amount of damage
Adhesive or inspection or sticky traps	Marking and recapture
Bait trap catch	Colonies of social insects
Narcotized collection	Transect and point sampling

Index